Self-Esteem

Dedicated to the memory of John Kevin Miller

SELF-ESTEEM

An American History

Ian Miller

polity

Copyright © Ian Miller 2024

The right of Ian Miller to be identified as Author of this Work has been asserted in accordance with the UK Copyright, Designs and Patents Act 1988.

First published in 2024 by Polity Press

Polity Press
65 Bridge Street
Cambridge CB2 1UR, UK

Polity Press
111 River Street
Hoboken, NJ 07030, USA

All rights reserved. Except for the quotation of short passages for the purpose of criticism and review, no part of this publication may be reproduced, stored in a retrieval system or transmitted, in any form or by any means, electronic, mechanical, photocopying, recording or otherwise, without the prior permission of the publisher.

ISBN-13: 978-1-5095-5940-4

A catalogue record for this book is available from the British Library.

Library of Congress Control Number: 2023934728

Typeset in 11.5 on 14 Adobe Garamond
by Fakenham Prepress Solutions, Fakenham, Norfolk NR21 8NL
Printed and bound in Great Britain by CPI Group (UK) Ltd, Croydon

The publisher has used its best endeavours to ensure that the URLs for external websites referred to in this book are correct and active at the time of going to press. However, the publisher has no responsibility for the websites and can make no guarantee that a site will remain live or that the content is or will remain appropriate.

Every effort has been made to trace all copyright holders, but if any have been overlooked the publisher will be pleased to include any necessary credits in any subsequent reprint or edition.

For further information on Polity, visit our website: politybooks.com

Contents

Acknowledgments · vi

Self-Esteem: An Introduction · 1
1 Happiness · 21
2 Race · 47
3 Sexuality · 76
4 Mothers · 103
5 School · 128
6 Teenagers · 151
7 Girls and Women · 167
8 A Self-Esteem Task Force · 197
9 The Twenty-first Century · 213

Notes · 225
Index · 269

Acknowledgments

This book took a number of years to write and was written alongside commencing a permanent academic role at Ulster University. I would like to thank those colleagues at Ulster University who have encouraged my career and to whom I am indebted for having kindly offered ongoing support. These include Justin Magee, Gerard Leavey, Conor Heffernan, Don MacRaild, and Kyle Hughes.

The project involved various trips to libraries and archives between 2018 and 2022, a period interrupted by the COVID pandemic. In particular, I would like to thank the staff at Wellcome Collection, the British Library, and Bishopsgate Library. The research benefited immensely from a number of exchange trips and academic talks. In November 2022, I was fortunate enough to spend a month visiting HEX (Academy of Finland Centre of Excellence in the History of Experiences), University of Tampere, Finland, and would like to acknowledge the support and interest among the excellent cluster of historians interested in emotions and experience at the Centre.

Aspects of this research were delivered at the Critical Perspectives on Mental Health seminar series (University of York), the European Association for the History of Human Sciences conference (Sigmund Freud Private University, Tempelhof Airport, Berlin), Ulster University's School of Psychology seminar series, the Canadian Society for the History of Medicine conference, the Institute for Medical Humanities annual conference (Durham University), the Southern Association for the History of Medicine and Science conference (Emory University) and the European University Institute (Florence). I wish to thank the organizers and participants at these talks for their insightful feedback, which has helped shape this book.

Throughout my career, I have been fortunate to work with an excellent group of postdoctoral and postgraduate researchers in medical history. I would like to acknowledge my PhD researchers Rebecca Watterson,

ACKNOWLEDGMENTS

Eugenie Scott, Hannah Brown, Rebecca Brown, and Michael Kinsella, as well as Rhianne Morgan. Also at Ulster University, I am particularly indebted to Ian Thatcher, whom I wish to acknowledge for his ongoing enthusiasm for this project and my career. Further afield, I am grateful for the ongoing support shown by Ida Milne and Bryce Evans, as well as numerous scholarly societies with which I became involved during the writing process, including the Society for the Social History of Medicine, History UK, and the Royal Irish Academy.

Most importantly, I wish to acknowledge Pauline, Katie, and Sarah Miller, and also Miriam Trevor. Finally, I am eternally grateful to Laura Garland for her ongoing support and love.

August 2023

Self-Esteem: An Introduction

In the twentieth century, the idea of self-esteem gained enormous influence. Psychologists published a staggering amount of research on the subject. Readers turned eagerly to self-help manuals hoping to confront their flailing self-image. Self-esteem experts came along, some from the psychology profession, some from outside. Self-esteem programs became increasingly popular in clinics, homes, and schools. Self-esteem became *the* way of understanding ourselves, our personalities, our interactions with others. After World War II, African Americans, feminists, and LGBTQ+ activists tried to collectively improve their self-esteem, pride, and self-respect, believing this to be a foundation for social equality and acceptance. Nowadays, we take for granted the idea that our self-esteem levels, whether these are high or low, profoundly influence how we feel as we go about our lives, and how well we succeed in life. Few of us think much about the nature of self-esteem, or where the idea came from, or how we apply it to our lives. But perhaps we should.

Some scientists believe that self-esteem levels are currently plummeting across the West, particularly in countries that tolerate higher levels of social inequality. Seeing low self-esteem as a social justice matter, they argue that this widespread emotional decline stems from harmful, but potentially rectifiable, socioeconomic and political conditions. Social epidemiologists Richard Wilkinson and Kate Pickett argue that:

> While we are all used to the idea that there are pollutants and carcinogens in the environment that have to be reduced in order to diminish the burden of physical disease, we are less used to the idea of tackling harmful emotional or psychological environments. Yet if the causes of heightened levels of social anxiety are the source of serious damage to social life and well-being, they surely warrant as much political and public attention as the air we breathe.[1]

In the West, we live in an "emotional environment" shaped by capitalism and individualism, although we tend to go about our lives without reflecting on the influence of these forces on our inner feelings. This book beckons readers to think about low self-esteem not as an innate, natural human feeling, but, on the contrary, as an inner feeling, a psychological state, influenced profoundly by the particular political, cultural, and socioeconomic structures, environments, and circumstances that envelop us.

Different cultures and societies understand, experience, and manage feelings in particular ways. Some attach tremendous importance to certain feelings (such as self-esteem) in ways that others don't. Historians of emotions argue that collective feelings also change over time as these cultures and societies transform.[2] As Western societies changed over time, the emotional environments embedded within them fluctuated and altered, bringing with them new collective emotional experiences. After World War II, self-esteem truly began to concern Westerners in ways that it previously hadn't. This book sets out to find out why, focusing on America. While researching this book, I initially intended to cross-compare America with other Western countries, but then decided to explore America alone. I made this decision upon discovering the wealth of material on self-esteem published in that country. Indeed, Americans have produced such a large amount of literature that it began to seem as though there was something particularly *American* about the self-esteem that demanded investigation.

Adopting a somewhat sweeping *longue durée* sociological approach, it's possible to think about America's recent fascination with self-esteem as having arrived as part of the West's historical transition away from small communities and religious authority toward modern urban life and democratic society. Older societies were less mobile, meaning that people were more familiar and intimate with one another. In those less competitive societies, it didn't seem necessary to constantly monitor what others were thinking of us, unlike how so many of us obsessively do today. Self-esteem and social standing came from personal and moral conduct, as well as behavior, and less from social status, wealth, or self-image.[3] The arrival of social modernity profoundly altered how we thought about, and assessed, ourselves; modern life changed Western consciousness, as our worldviews were restructured in new and unfamiliar ways. The West was made modern not only by revolution, industrialization, and the Enlightenment, but also by the advent of a "society of strangers." Rapid and sustained population growth increased mobility over expansive distances, concentrated people in large cities, separated individuals and families from each other, and broke the strong communal ties that once characterized premodern societies.[4]

Our mentalities changed too. Modernization, with its disruption of social, familial, and geographic roots, brought significant emotional consequences.[5] Urbanization cut people off from one another, heightening feelings of isolation and unfamiliarity with others; capitalism encouraged Westerners to be competitive rather than communal.[6] The industrial, urban-based societies that developed from the late eighteenth century offered choice, freedom, independence, potentially even happiness, but they also raised everyone's potential to feel alone, alienated, uncompetitive, and inferior. While encouraging success, happiness, and competitiveness, modernity left many of us feeling isolated, inferior, doubtful, and lonely.[7] As one critic outlines: "The lonely isolated person is endemic to neoliberal capitalism; in this sense, the system itself is unhealthy and breeds problems and unhappiness." Others believe that the historical shift toward urban living produced specific psychobiological changes in our brains and psyche, creating the mental health concerns that are rampant in today's urbanized cultures.[8]

Nowadays, those of us living in the West are less likely than our predecessors to die of material poverty or hunger (although, of course, such problems still exist), but we do seem to suffer more widely from depression, anxiety, eating disorders, and bipolar conditions. Some researchers attribute this to the alienating circumstances of modern social life.[9] The increase in diagnoses of these mental health complaints is especially prevalent in industrialized Western countries. It was not until the 1990s that people living in communist countries – for example, China – started thinking of themselves in psychological terms. Before then, China had no psychotherapeutic culture. The country's collectivist politics mitigated against anything suspiciously reminiscent of Western individualism.[10]

The post-World War II period was a time when capitalism, immigration, loss of community, and consumerism accelerated a Western emotional epidemic of isolation and dissatisfaction; it was an era when feelings of emptiness, low self-esteem, and despair began to flourish. The so-called "empty self" emerged – someone who felt a desperate constant yearning to be loved, soothed, and made whole; who sought to fill an inner emptiness, to develop confidence, and to make contact with others.[11] Again, this was a consequence of social change. After the war, many Westerners, Americans in particular, moved to the suburbs. They lived alongside people whom they did not know, and often felt estranged and distant from those around them, struggling to make new connections. Learning to get along with unfamiliar people, developing enough strength of personality to do so, all on top of being socially and economically successful, became essential, although this ultimately proved difficult for many.

The "other-directed person" began to dominate American society: a person who could identify themselves only with reference to others, constantly comparing what they earned, owned, consumed, and believed in. At the same time, a thriving market grew for books on positive thinking. The confident, socially extroverted personality became the new American ideal, and remains so today.[12] Nonetheless, it is striking how little attention self-help, including associated ideas such as self-esteem, has received from historians given its enormous influence.[13]

Postwar affluence allowed most Americans to stop worrying about basic physical needs such as hunger and shelter. In a new era of abundance, time was freed up to focus on non-essential, emotional needs.[14] In the 1950s, humanistic psychologist Abraham Maslow influentially defined these as esteem, belongingness, and love.[15] Forty years later, psychologist Julia A. Boyd recalled:

> Both my mother and father were raised by deeply religious Southern parents. From their parents my parents learned the basics of life: the Bible got you into heaven, strong discipline taught you respect, and common sense kept you alive. Self-esteem wasn't the issue for my parents or their parents. Survival was their primary goal ... The concept of loving one's self was a foreign concept to my parents, because it was a foreign concept to their parents ... The majority of us had parents who were good at fulfilling our physical needs.[16]

But we didn't become happier. In light of rising psychiatric diagnoses, many people speculate that we are living through an "age of anxiety" or a catastrophic "mental health epidemic." Writing in the context of the 2020 COVID pandemic and its detrimental impact on our mental wellbeing, sociologist Jason Schnittker argued: "If there ever were an age of anxiety, it seems to be now. As a category, anxiety disorders are the most common psychiatric disorders in the United States." Like other sociologists, Schnittker attributes this to "the accumulation of slow-moving transformations, spanning eras and generations."[17] As societies modernized, anxiety followed. Alain Ehrenberg, in his study of rising depression diagnoses, similarly surmises that the mindsets and attitudes characteristic of Western modernity drove a portentous epidemiological change. Historical emancipation from religious authority "made us persons without guides, progressively placing us in a situation in which we need to judge what surrounds us by ourselves and construct our own reference points." Ultimately, Ehrenberg portrays depression as a pathology of a (Western) culture that privileges responsibility and initiative but ends up creating only a "pathology of inadequacy."[18] Similarly, historians of emotions frame loneliness as a "disease of civilization," a feeling that became commoner from the nineteenth century largely due to industrialization, modernization, and the privileging of individual over collective needs.[19]

Of course, we should be cautious of taking health statistics at face value. The actual prevalence of (often vaguely defined) emotional states is impossible to know, and just as many Americans claim to feel happy as they do sad or depressed. Doctors tend only to see the most advanced, serious cases; diagnoses are shaped by a broad range of factors that change over time. And just because societies begin talking more about certain emotional states (such as low self-esteem) doesn't necessarily mean that those conditions are actually increasing, but rather that societies become more concerned about them at particular times. Nonetheless, it still seems feasible that low self-esteem is yet another emotional malaise connected to modern ways of living. Hierarchical, competitive societies lead too many people to view themselves as "lower," "inferior," and "less admirable." Low self-esteem tends not to be categorized as a sickness per se, but as an affective state with enormous potential to lead to more severe mental health and psychosocial problems, and "a cancer in the midst of our social life."[20]

Self-esteem's critics

Interest in self-esteem surged after World War II largely because of psychology's more general growing influence. Only in the twentieth century did Westerners start thinking of themselves comprehensively in psychological terms, and with reference to ideas about emotional wellbeing. Psychology as we know it today developed from the late nineteenth century, when Sigmund Freud published his influential psychoanalytic theories and American psychologist William James transformed psychology from its philosophical roots into a therapeutic endeavor.[21] Psychology's development took many turns and directions. Nonetheless, and particularly after the war, psychological thought, practice, and expertise impacted hugely on American society.[22] We became "psychological subjects," viewing ourselves and the world around us in affective terms.[23] We started to think about ourselves in relation to our emotions and feelings, and less as social, economic, or political beings (although all of these were deeply intertwined with the psychological). Psychology played a major role in "inventing our selves."[24] According to social theorist Nikolas Rose, "human beings have come to understand and relate to themselves as psychological beings, to interrogate and narrate themselves in terms of a psychological inner life that holds the secrets of their identity, which they are to discover and fulfil."[25] In this context, the idea of self-esteem took root and flourished.

While writing this book, I often found myself guilty of referring to self-esteem in imprecise ways that lacked rigorous definition, and using the term interchangeably with related concepts. For clarification, self-esteem is the term psychologists use for how we value and perceive ourselves. Self-confidence is a related term, referring to our belief in ourselves and our abilities. However, it is possible to be confident about abilities while suffering low self-esteem. Self-image refers more specifically to the ways in which we see ourselves, while self-concept refers to three connected components: self-image, self-esteem, and the ideal self. In this book I also discuss at length related terms such as pride – gay pride, Black pride, for example. Feeling inferior, or ashamed of oneself can be loosely defined as the opposite of feeling proud. While inferiority complexes operate on a personal, individual level, activist groups have often used the concept to describe a collective feeling of being made to feel inferior by dominant social groups, and upheld increasing self-esteem as the solution.

Psychology and religion have had a precarious relationship. Pride is one of Christianity's seven deadly sins.[26] Although much self-esteem advice was initially written from a religious perspective, by the 1970s the church felt far more threatened by psychologists. Secular psychological expertise was rapidly supplanting Christianity's traditional role as the source of social guidance and authority. That decade saw an unprecedented expansion of the self-help book market, which accounted for more than 15 percent of bestselling books in America.[27] A religious language of suffering now became supplanted with a diagnostic one.[28]

Traditionally, Christianity called for a denial of the self, but, as sociologist James L. Nolan Jr. argues, the self – largely thanks to psychology – is no longer to be surrendered, denied, or sacrificed. Instead, it exists to be esteemed, actualized, unfettered, and affirmed.[29] Broadly speaking (and there were many exceptions within Christianity), traditional religion encouraged conformity and preached punishment. The self-esteem movement promised the opposite: liberation from all constraints.[30] In the 1970s, Paul C. Vitz portrayed psychology as a cult of the self. Self-worship was replacing God worship, but Vitz believed that God provided more than enough love and should decide how happy we feel. Self-help amounted to atheism.[31] These debates continue in the twenty-first century.[32] (Of course, it's possible to accuse religion of causing harm. In his popular book *Selfie*, Will Storr insists that his Catholic upbringing instilled within him an overbearing sense of guilt and shame inculcated by the Church's fetishization of low self-esteem and self-hatred.[33])

However, even within psychological circles, the concept of self-esteem has been regularly questioned for its uncertainty and vagueness. Self-esteem is a firmly established psychological category, one considered to be of utmost importance, but confusion prevails about its precise nature. No matter how much research is undertaken (and there is a lot of it), the idea of self-esteem remains ambiguous, lacking precise definition. The term self-esteem is used widely but uncritically.[34] In 1979, social psychologist Morris Rosenberg, a leading expert, admitted: "It is somewhat astonishing to think that after decades of theory and research on the self-concept, investigators are as far as ever from agreeing on what it is or what it includes." In 1981, another prominent expert, Stanley Coopersmith commenced his book, *The Antecedents of Self-Esteem*, with a prefixed warning that "so little is known about the conditions and experiences that enhance or lessen self-esteem," adding that "there is not one single theoretical context in which self-esteem can be considered without accepting a number of vague and often unrelated assumptions."[35] In the twenty-first century, researchers still describe the concept of self-esteem as "vague" and "elusive."[36] But, nevertheless, the basic idea of self-esteem thrives.

Some social psychologists describe self-esteem as a modern "social imaginary": a broad understanding of the world embraced by large numbers of people that, at some point, became so taken for granted that we all forgot that this way of constructing things was not necessarily the only way.[37] Self-esteem became so central to American culture that, nowadays, few people ask where the idea came from or reflect upon how it is used. Self-esteem, as an idea, is now "naturalized." To borrow from sociologist Frank Furedi, "so widespread are deliberations about self-esteem that it is easy to overlook the fact that the problems associated with it are of relatively recent invention." Intriguingly, Furedi also states: "The self-esteem deficit is a cultural myth that is continually promoted by its advocates."[38]

"Imaginaries." "Myths." One might conclude that self-esteem isn't real after all, that it's been fabricated, for whatever reasons, by psychologists. Certainly, critics accuse psychologists of labeling behaviors as illness and normal emotions as inner sickness. At some point in therapeutic history, sadness became depression and sad people turned into patients requiring therapeutic and psychopharmaceutical intervention. In their pursuit of happiness, Westerners started to feel unduly afraid of natural feelings such as fear and sadness.[39] The most determined critic was Hungarian psychiatrist Thomas Szasz who, from the 1960s, dismissed mental health diagnoses as a "myth" lacking objective reality. He argued that psychiatric diagnosis captured behaviors that were deemed antisocial.[40] This pathologizing of natural emotions unnecessarily encouraged people to feel abnormal, depressed, even traumatized.[41] One example of the fluidity and subjectivity of mental health diagnosis is the formal categorization by the American Psychiatric Association, which remained in place until 1973, of homosexuality as a psychological disorder requiring therapy.[42]

Psychology undoubtedly creates its own subject matter, but I would disagree with perspectives that portray this simply as a top-down process.[43] There is no simple distinction between "constructed" and "real." Ian Hacking discusses the "looping effect" concept, whereby psychologists and social scientists initially develop new knowledge about a type of person under their observation. Hacking uses the example of the "criminal" or "homosexual," but the "low self-esteem person" also suits our needs here. Once new knowledge is created of this person and becomes known to the person classified, it changes the way these individuals behave and think about themselves. This changing behavior then loops back forcing experts to reclassify the knowledge they had originally assumed. Hence, understandings, and the behavior, of low self-esteem individuals are unstable, and can change over time and in different places. The personality type under examination is, using Hacking's term, a "moving target."[44] In that sense, we interact with new disease classifications. People classified in certain ways grow into the ways described, but these descriptions are constantly revised.[45] For example, even if we suggest that "race" isn't real, the effects of dividing people into different races undoubtedly are tangible.

I would argue though that these experts have played a pivotal role in establishing ideas about which of our emotional states are "good" or "bad," "positive" or "negative," "healthy" or "unhealthy." We believe that low self-esteem isn't "normal" or "healthy" largely because psychologists have told us so. It threatens the pursuit of happiness, the West's idealized emotional state.[46] However, it would be going too far to dismiss the concept of self-esteem entirely. Undoubtedly, the psy-sciences carry tendencies to advance fictional images of the idealized, emotionally healthy human being, one brimming with self-esteem, confidence, and happiness, a central theme of this book. Storr describes this perfect person as an "extroverted, slim, beautiful, individualistic, optimistic, hard-working, socially aware yet high self-esteeming global citizen with entrepreneurial guile and a selfie camera."[47] But we risk overlooking real emotional suffering by denying, or not taking seriously enough, low self-esteem as a significant modern problem.

Moreover, we need to critically interrogate how American society has experienced and manage self-esteem, even if doubts exist about the concept's veracity. For all we know, self-esteem might well be merely another example of meaningless "psychobabble," but the point still stands that self-esteem evolved into arguably *the* most influential psychological concept, one used to explain all manner of personal, emotional, and social problems. All of us now experience our lives with reference to ideas about our self-esteem. We need a better understanding of its history, politics, and implications, and we need to take seriously the vast numbers of people who willingly absorbed the idea, imbuing it with profound meaning, and who experienced negative self-image as fundamentally disabling.

Narcissistic modernity?

In the 1970s, fears emerged that Americans had created a menacing new world characterized by a dangerously self-absorbed culture. Self-obsession was discussed as the decade's American plague. The debates continued well into the 2000s, when some social critics insisted that modern Americans had lost any sense of obligation beyond the narrow compass of their self-absorbed lives. Twenty-first-century American youngsters seem noticeably more narcissistic than their predecessors, the Gen Xers and the Baby Boomers. Such literature undoubtedly caricatures young Americans as self-centered, narcissistic egomaniacs, while disdainfully ignoring their many positive attributes, and failing to recognize that narcissism had its complexities and contradictions.[48]

Rather than offering yet another study of narcissism, this book draws attention to the harmful emotions that often loom beneath fragile masks of egotism. It explores self-esteem as a feeling or emotion resting somewhere below the narcissistic scale, a psychological state potentially healthier than narcissism.[49] The popularity of self-help courses and self-esteem books suggests that a buoyant market exists of people who struggle to feel happy and confident. "Self-obsession" has not always led to excessive, unwarranted self-love. Take this quote from Storr's aforementioned book in which he laments Western cultural obsession with achieving perfection and securing positive approval from others to bolster self-esteem, a trend he believes has intensified greatly since his 1980s youth. He agrees that, collectively, our mentalities changed in recent decades. Now, hazardous psychological agents exist in our emotional environment because Western society transformed for the worse:

> We're living in an age of perfectionism, and perfectionism is the idea that kills. Whether its social media or pressure to be the impossibly "perfect" twenty-first-century iterations of ourselves, or pressure to have the perfect body, or pressure to be successful in our careers, or any of the other myriad ways in which we place overly high expectations on ourselves and other people, we're creating a psychological environment that's toxic.[50]

One might read from this not so much an overriding narcissistic love of oneself as an underlying deficit of faith in oneself that is pervasive in modern Western society. This raises questions about whether narcissism's critics have failed to fully acknowledge the realities of sadness, stress, worry, anxiety, or depression across all generations, and particularly among the young, a group that forms much of the focus of this book. Storr, when reflecting upon self-obsession, developed instead a narrative of inner sadness, of inabilities to live up to social pressures that ultimately harmed emotional wellbeing. We should take these matters more seriously.

At this point there emerges a question of fundamental importance to this study: if self-esteem levels are shaped by political, cultural, and socioeconomic forces, why have self-help experts consistently advised us to self-improve and work on our inner mentalities, instead of actively pursuing sociopolitical change? This paradox points to Western psychology's potential to discourage reflection upon the world residing beyond the confines of the mind, to individualize and depoliticize life's circumstances, and to advocate an intent focus on inner emotional work that conceals the structural roots of our problems. In the process of replacing religious authority with self-referential forms of governmentality, we developed a therapeutic culture hellbent on personal transformation without consideration of structural forces and toxic emotional environments. This is dysfunctional. A purely psychological view of human difficulties mystifies social reality.[51] It fails to tackle head-on the emotional misery that many people consider characteristic of capitalism.

The rise of therapeutic culture was first noted in the 1960s and, since then, its critics have lambasted a decline of communal values, increased personal isolation, and heightened reliance on mental health experts.[52] The term "therapeutic state," coined by Szasz, refers to political strategies that encourage individuals to internalize and personally resolve life's problems.[53] Typically, these strategies direct blame for social problems toward individuals and then entice them to improve through therapeutic self-help means. The myth of meritocracy, which persuades us to believe that success arises from hard work, only adds to the anxiety, despair, and powerlessness felt by those encouraged to blame themselves for life's misfortunes.[54]

A conservative backlash occurred from the 1970s that rallied against the counterculture. In the 1980s, Ronald Reagan's presidency further weakened fights for social justice in part by stressing the importance of individual endeavor rather than government action as a way of fostering happiness, character, and virtue. Therapy was a vehicle that conveyed Reaganite notions of individual responsibility. In the West during the Cold War years, overtly authoritarian forms of governance were out of the question. Governance was achieved by educating citizens, and by providing the freedom to take advice on board (such as self-help) Neoliberal governance is coercive in the sense that it encourages us to govern ourselves while ostensibly offering us freedom. Capitalist mentalities hold each of us responsible and accountable for our actions and wellbeing by presenting life's problems as fundamentally personal, not structural, in origin. In their reluctance to fully address social issues, deliberately or otherwise, capitalism and Western psychology share common ground.[55]

Similar accusations are currently leveled at the fashionable mindfulness movement, also criticized for a naive blindness to structural inequalities. Some slam the "mindfulness revolution" as banal capitalist spirituality that willfully avoids sociopolitical transformation. Mindfulness educators refuse to recognize and resist the inequitable, individualistic, market-based society that creates widespread inner distress. Wellness has been presented as a syndrome itself that reduces dreams of social change to individual transformation. Suspiciously, mindfulness techniques have been harnessed by the corporate world. All of this encourages us to adjust to the status quo instead of rallying against it. Downplaying socioeconomic and political context dissuades people from thinking about and challenging social inequity.[56] Skeptic Ronald E. Purser argues that "anything that offers success in our unjust society without trying to change it is not revolutionary. It just helps people cope." Purser describes mindfulness as a "tool of self-discipline disguised as self-help." For him, "the best we can hope for is palliative care in a neoliberal nightmare, adapting to the brutal forces that afflict us. This is a morally and spiritually bankrupt way to imagine human lives."[57] Another variant is the idea of "grit," which emphasizes that success depends upon what how we respond psychologically when we fail.[58] Resilience is yet another concept with a surprisingly long and intriguing history. It became widely used in psychology from around the 1960s. It was popularized by being used to describe soldiers who returned from conflict seemingly free from conditions such as PTSD. It is currently central to twenty-first-century ideas about positive psychology and mindfulness.[59]

According to these accounts, psychological knowledge equips us with emotional stamina to endure, but never challenge, adverse circumstances and injustices, being in that sense deeply conservative. Self-image improvement leaves intact harmful emotional environments. Turning to self-help alleviates feelings of despair, anxiety, and low self-worth, but overlooks the broader external roots of negative inner feelings.[60] This has been described as seeking "biographical solutions for structural crises."[61] Critic Carl Cederström writes that "the optimistic notion that everyone could actualize their inner potential has now turned into a cruel and menacing doctrine, strategically employed to sustain and normalize the structural inequalities in contemporary capitalism."[62] In the seemingly helpful, innocuous self-help book, a new apolitical self is forged.[63] The concept of self-esteem, its critics insist, tries to cast us as isolated individuals rather than interconnected parts of sociopolitical units.

In this book, I trace how, gradually over time, the self-esteem movement urged us to forget the broader external determinants of our more destructive emotions. It outlines how negative emotions became depoliticized, isolated into personal, individual challenges that we alone, all by ourselves, needed to overcome. Working on one's self-esteem became marketed as the *solution* to life's problems, an approach devoid of sociopolitical context and woefully divorced from social realities. In reality, low self-esteem is probably a telling symptom of the modern West's inherent problems. (This is not to say that American conservative rhetoric is entirely uncritical of the self-esteem movement. Indeed, conservative culture is closely linked to the aforementioned religious groups that view self-help with suspicion, and scorn participation trophies and other schemes that have become increasingly common in American school education. Nonetheless, and perhaps ironically, conservative religious critics and self-esteem advocates share common agendas in advocating reactionary, rather than socially revolutionary, political agendas.)

A revealing example of how low self-esteem has been portrayed as the cause of social ills (rather than vice versa) is found in a foreword to Sandy McDaniel and Peggy Bielen's *Project Self-Esteem: A Parent Involvement Program for Elementary-Age Children* (1986), written by influential American author and motivation speaker Jack Canfield:

> As an educator and facilitator deeply concerned with self-esteem, I have conducted intensive self-esteem seminars for over 26,000 people. It is an interesting commentary on our educational system that hundreds of thousands of adults find the need to spend $250 to $500 for weekend workshops to learn to "feel better about themselves." We live in a country with many people who don't like themselves. Riddled with guilt, fear and psychological ignorance, many people live in chronic emotional and relational pain, some of it acute and some of it what we have grown to accept as "normal."
>
> In attempts to numb out this pain, millions turn to alcohol, drugs, sex, workaholism, television addiction, overeating, smoking, and withdrawal as a means of coping; anything so as not to experience the pain and hopelessness that accompanies their loneliness, separation, and rejection. According to many recent polls, four out of five Americans have low self-esteem. What's worse, the results of low self-esteem show up everywhere – in low productivity, academic underachievement, illness, depression, violent crimes, child abuse, incest, and rape – all of which have been on the rise recently. With statistics like these, it's easy to feel overwhelmed.
>
> But don't despair. There is really a lot of good news around. Answers to the problems, solutions that address their core cause, are starting to show up across the country.[64]

One answer, of course, is to work on one's self-esteem. While we might expect self-help advice to be cheerful, hopeful, and supportive, the emphasis on rape, incest, child abuse, and acute mental illness appears unexpectedly. Their incorporation in self-help books is morally persuasive, placing obligations on readers to work on their own self-esteem, and also on that of their children and pupils. (McDaniel and Bielen's book was aimed at both teachers and parents.) Improving one's self-esteem is certainly optional. Readers of *Project Self-Esteem* could choose to discard the solutions on offer if they wished, but ignoring them altogether had potentially dire moral implications for the young. Far from optimistic, Canfield's foreword renders visible the persuasive undertones that are rife in much self-help advice.

Jack Canfield saw little need to promote structural change. Why bother campaigning for new laws to protect women against rape, prevent child abuse, and improve educational facilities? Why waste time considering the socioeconomic conditions that breed violence, abuse, rape, and substance misuse? In the 1980s, boosting one's self-esteem was all that was needed to make these problems magically disappear. At least that was the logic of capitalist-driven interpretations of self-esteem advice. From the 1990s, Canfield acquired huge popularity as a motivational speaker and self-styled godfather of self-esteem with his *Chicken Soup* series of books.

We can certainly opt out of working on our self-esteem. But when experts warn us that the consequences include children growing up delinquent, relationships suffering, and an increased likelihood of turning to drugs and alcohol, does it genuinely leave us with much choice? This is a manner of governing that relies not on institutions or state power or violence, but on securing citizens' voluntary compliance.[65] To speak of this governance is not to imply social control directed by governments or the medical profession, but to recognize the existence of morally persuasive forms of power that urge us to govern ourselves in particular ways and align our behaviors with core capitalist values.

Activism and self-esteem

But then again, is the therapeutic state really so pervasive and effective at quietening political activism? Some political scientists warn that the therapeutic state "looms as a menacing Leviathan." Other critics present therapeutic culture as ubiquitously successful in directing attention away from structural inequalities.[66] Philip Cushman, for instance, writes:

> If we cannot entertain the realistic possibility that political structures can be the cause of personal, psychological distress, then we cannot notice their impact, we cannot study them, we cannot face their consequences, we cannot mobilize to make structural changes, and we will have few ideas about what changes to make. We will become politically incompetent. Perhaps we already are.[67]

In my view, these accounts of the rise of neoliberal emotional management tell only one side of a more intricate story. Some researchers query assumptions about therapeutic culture's entirely insidious nature, arguing that there are tales to tell beyond it being a monolithic "harbinger of decline" and "social control." Therapeutic culture can foster a greater recognition of pain and suffering. Rather than being solely self-indulgent, ideas about self-help, self-esteem, and self-respect have been used to assert individual rights to bodily autonomy, emotional wellbeing, and personal safety. Therapeutic culture offers opportunities to express and expose experiences among the least powerful groups through a language of emotional pain that persuasively articulates injuries to the self and might even help redress psychological damage. Therapeutic culture can therefore be multidimensional.[68]

In their 2015 book, Timothy Aubry and Trysh Travis warn against casting therapeutic culture aside as entirely wrong-headed and pernicious. For them, "therapeutic culture's complexity, its capacity to serve seemingly incongruous ends, has been overlooked by scholars and critics of the therapeutic." All that materializes is a "cultural history of the therapeutic that is conceptually thin, ideologically blinkered and, unsurprisingly, not that useful."[69] Aubry and Travis urge us to better understand the vagaries of therapeutic culture's evolution, the nuances of its aesthetics, the diverse implications of its politics, and the unexpected political and social functions it may have performed.

And that is where this book departs from the critics outlined above. I concur that self-help bolsters capitalist agendas, but argue that marginalized groups, particularly African American, LGBTQ+, and feminist activists, all identify self-esteem as a pressing problem for their communities. They link low self-esteem to social injustice, develop ways of articulating negatively experienced emotions, and actively seek to resolve psychological problems as part of their emancipation. These groups are deeply cognizant of the relationship between the structural and the emotional. Radicalism and therapy are not always entirely at odds, and many people have viewed inner self-work as an effective stepping-stone toward social change.[70]

Scholars of race, gender, and sexuality all argue that low self-esteem matters considerably to disadvantaged communities. In 2003, Black feminist author bell hooks lamented that cripplingly low self-esteem was epidemic among African Americans. This was more than a simple reflection of heightened psychological distress across America more generally. Additional to modernity's general emotional stress, and unlike whites, Blacks contended with racism, school and workplace discrimination, and white supremacist culture, among a plethora of other challenges.[71] Since the 1990s, clinical psychologist Walt Odets has argued that LGBTQ+ communities have internalized the stigma directed toward them. Despite an optimistic decade of gay pride and liberation in the 1970s, the 1980s AIDS crisis, which disproportionately affected gay men, hardly produced a gratifying self-image or improved straight peoples' opinions. At worst, it left behind a legacy of grief, shame, and trauma among its survivors.[72] The self-esteem of girls and women has been fiercely debated. In 1992, feminist Gloria Steinem published her *Revolution from Within: A Book of Self-Esteem*. However, feminists still inconclusively debate the value of individualistic self-help strategies for collective action.[73] Critical studies of therapeutic culture, although sophisticated, insightful, and nuanced, tend to ignore the impacts of broader sociological forces on the marginalized. Bringing these groups into the fore, as this book does, will help develop a more rounded understanding of therapeutic culture.

In the 2020s, some attention has been paid to this hidden history with reference to self-care. Interest blossomed during the COVID pandemic. A spate of articles appeared that called attention to self-care's radical origins. In *Slate* magazine, Aisha Harris argued that self-care first surfaced in the 1960s in radical Black Power and feminist political movements which insisted that taking control over one's health (physical and emotional) was a radical act that implicitly challenged the lack of value granted to certain groups in patriarchal, white American culture. Here lay the foundations of twenty-first-century initiatives such as self-care and wellness.[74] This was echoed by other online articles.[75] *Teen*

Vogue magazine described the foundational role of the Black Panthers. At the time, George Floyd had recently been murdered and the Black Lives Matter movement was resurging.[76] That same year, a BBC series similarly traced self-care ideas back to the Black Panther Party.[77]

I agree that, from the 1960s, activists, on behalf of the marginalized, used self-help techniques to advance agendas of collective improvement, with self-esteem being crucial to their activism. Improving one's self-image was an intrepid first step toward collectively transforming the psyche of marginalized groups. Women, African Americans, LGBTQ+: all these communities actively, and publicly, strove to replace shame with pride. Thus emerged Black pride, gay pride, and so on. However, this book adds sophistication to the current historical narrative. For me, this presumption that self-care had its roots in late 1960s and early 1970s activism presents problems. Was the path between the 1960s and 2020s really so linear? What happened to self-care, or its antecedents, in the fifty-year period left unspoken about? Were radical Black and feminist self-care campaigns really as effective as nostalgically presumed? What opposition emerged toward them? And what of the LGBTQ+ activists who upheld self-esteem and pride as core collective goals, but whose efforts are rarely recalled?

Ultimately, this book portrays the radical capturing of self-help in the 1960s and 1970s, with its rallying call for self-esteem, self-love, and group pride as projects with considerable historical intricacy. For a start, as I demonstrate, feminists didn't engage fully with psychological approaches such as self-esteem until the 1980s, not the late 1960s, partly due to lingering skepticism about whether the psy-sciences were supportive of patriarchal oppression. Recent self-care histories also overlook the fact that, despite Black Power's efforts, low self-esteem persists as a major problem in African American communities, suggesting limitations to Black pride. Also, tensions existed between, and even within, the various marginalized groups. Historically, many white gay men held racist views toward Black gay men; men in the Black Power movement held conservative attitudes toward Black women; and Black attitudes toward Black gays could be unfavorable, to say the least.[78] The 2020 US miniseries *Mrs. America*, which dramatized the 1970s battles between feminists such as Gloria Steinem and Betty Friedan with conservative women led by Phyllis Schlafly, foregrounded the complex intersections between gender, race, and sexuality, even within seemingly socially transformative strands of leftwing feminism. Critics still accuse mainstream feminism for catering first and foremost to white, middle-class Western women, while excluding women from other backgrounds.[79] Intersectionality is important too. Some self-esteem books have been aimed at, for example, Black women, who feature in two groups subject to discrimination.[80] (There are a number of minority groups that I do not cover exhaustively, partly through reasons of space, but also because they didn't figure as prominently in my source material. These include Hispanics, Asian-Americans, Latinos, and Native Americans.)

To fully understand the history of self-esteem, we also need to explore the intersection of radical and capitalist-oriented self-help approaches. A key argument in this book is that, from the 1970s, self-esteem initiatives implemented in schools, as well as those advocated in popular self-help books, entirely ignored the strident work being done by activists trying to manage self-esteem and instead blamed social structures for worsening emotional wellbeing. Self-esteem advice, penned outside activist circles, promoted individual change while paying lip service at best to the external world and surrounding "emotional environments." It generalized by ignoring the unique emotional needs of marginalized groups, speaking instead to a white, middle-class (even, initially, male) audience, whether that be parents, youngsters, or adults. A one-size-fits-all approach emerged, in keeping with the broader backlash against leftwing activism, which addressed the idealized, "normal" American, an individual who was not Black, gay, or working class. American definitions of a "normal" human being implicitly excluded both nonwhites and non-straights, while rendering women inferior to men.[81]

Hence, micro-practices such as reading a self-help book or attending a self-esteem class tacitly maintained the status quo simply by having nothing at all to say about the specific emotional needs of minority groups. What we might term "capitalist self-esteem advice" was silently anti-diversity, but also hugely influential in America. The fluidity and vagueness of the self-esteem concept, open to interpretation, difficult to define precisely, allowed the idea to appeal simultaneously to differing, even competing, political agendas. The neoliberal and emancipatory aspects of self-esteem coexisted in uneasy tension with each other.

African American, LGBTQ+, and feminist activists harnessed self-esteem to promote freedom, liberation, and social equality, but did so in the face of (sometimes in response to) contradictory conservative strategies intent on concealing social inequalities and discrimination. Exploring how minority groups spoke out about their experiences in a language based on therapy and emotions, and in the face of modernity, capitalism, and the conservative backlash, allows us to think more carefully about therapeutic culture's history. By foregrounding activist groups in self-esteem's history, a nuanced picture develops that moves beyond lamenting the emotional misery of capitalist modernity, one that is inclusive and sympathetic to the diversity of emotional experience and the multiple meanings and uses of therapeutic culture. Delving more fully into the intricacies of self-esteem's politics reveals its curious capacity to simultaneously support, explicitly or implicitly, the politics of both liberation and conservatism. This shouldn't surprise us too much. While most readers will no doubt be familiar with ideas of gay, Black, or female pride, it must be remembered that the term "pride" is also harnessed by rightwing activists, further demonstrating the concept's fluidity. Groups such as the Proud Boys have recently emerged, a far-right, neo-fascist, white nationalist, and exclusively male organization, demonstrating further how concepts such as pride can be used simultaneously for conflicting purposes.

Before we enter into a fuller discussion of the complex politics of pride, it's necessary to look further back to the nineteenth and early twentieth centuries. My opening chapter invites readers to think about how and why self-esteem, and associated ideas such as happiness and positive thinking, first gained such broad appeal in America. It will take readers through the decades in which self-esteem (and ideas such as inferiority complex) first began to permeate both professional psychological thought and American popular culture, asking why self-esteem, happiness, and self-confidence became so important to Americans.

ONE

Happiness

In twentieth-century America, a brand-new image of the self-confident American individual was forged, one brimming with self-esteem, success, and positivity. This was a caricature, unobtainable to many; a new ideal that failed to acknowledge the sheer complexity and diversity of people's emotional life. Problematically, as psychologists produced models of happy individuals filled with self-esteem, they inadvertently rendered those who felt less confident about themselves as potentially ill and vulnerable, as if their life problems were somehow their own fault. In this development, we can observe the burgeoning influence of ideas that inner emotional work was essential to successfully tackling life's many tribulations.

The idea that we should constantly enhance our self-esteem to truly feel happy and content is relatively recent. Self-esteem is by no means a new concept, and nor is happiness, but experts and the public began to think about these ideas differently during the twentieth century. The Victorians used to believe that anyone could become successful by developing their characters and work habits. Perseverance was crucial; failures, illness, or poverty all needed to be overcome. Ambition was crucial too, although authors such as Charles Dickens warned against envy, excessive competitiveness, and selfishness. Hard-working individuals deserved their rewards, so long as they didn't take opportunities away from anyone else. However, as in current self-help approaches, uneasy tensions arose between pursuing individual and collective needs. Victorian self-help narratives taught that individuals could succeed together, working selflessly for the collective benefit. Then and now, the self-help genre largely elided systemic causes of unhappiness, while emphasizing personal responsibility in ways that remain familiar today.[1]

What made twentieth-century ideas about self-esteem (and self-help) so novel was their intent focus on personality and emotions, and not just on character and material gain. Over the course of the century,

American culture imbued self-confident personalities with considerable social capital. As in the Victorian era, fulfilling destinies still required hard work, but this involved inner work rather than, say, economic work. With proper affective training, self-help readers were promised full potential to gain control over their life and, importantly, feel happy.[2]

Happiness became the ultimate goal of the self-esteem movement, an emotion that served as the basis for a fulfilling life regardless of one's material circumstances; but the idea of happiness needs to be deconstructed. Since the twentieth century, Americans have upheld happiness as the optimal emotion, a feeling that everyone should always strive toward. Nowadays, few of us question the desirability of happiness. Why would anyone not want to be happy, after all? But happiness hasn't always been a key human aspiration. Nineteenth-century Methodists saw humans as inherently wicked and sinful, as people put on Earth to atone before ultimately dying and proceeding more happily to heaven. They favored suffering and misery as the preferred emotional experience, which explains their harsh treatment of Sunday School pupils.[3] This illustrative example highlights how different societies and communities hold particular ideas about how ideal feelings.[4]

Anthropologists add that happiness is not valued in all non-Western cultures. Some are even openly averse to it.[5] The same applies to self-esteem. Cultures exist that uphold self-criticism as the route to success. Some Eastern cultures don't even have a word for self-esteem, nor do they fully recognize the concept.[6] In light of this, we can begin to think about, and problematize, the happiness (and self-esteem) ideal as culturally specific to postwar Western modernity. It is also time-specific. The "happiness fantasy" first emerged in America in the 1920s. In the decades following World War II, happiness ideals fed into ideas about self-actualization (reaching our full potential as a human being) and, later on, became central to capitalist mentalities.[7] Self-esteem acquired particular importance from the 1970s onward, the same period when capitalism and consumerism truly began to dominate American culture.

Self-esteem and happiness were deeply entwined ideas. Ironically though, this new cultural emphasis on developing happy personalities, abundant with self-confidence, only deepened many people's personal anxieties. Presumably, few Americans felt fully able to transform into this glorified, exalted, but ultimately hypothetical, human being.[8] In reality, they experienced a fuller range of emotions rather than one-dimensional happiness. Immense pressure emerged to constantly monitor, regulate, and adjust our personalities, alongside a cultural obsession with how others perceive us. The consequences were sometimes negative. As Will Storr recollected about his childhood in the 1980s:

> What was remarkable was the extent to which I'd internalized these ideas. It was everywhere, during my teenage years, and my self had accepted Vasco's [John Vasconcellos] fantasy and used it, magpie like, to build my narrative identity. For reasons I'd suspected were to do with my family environment, and Catholicism, I was a person with low self-esteem and one of my missions in life was to raise it.
>
> But, more damaging than that, I'd also absorbed my culture's perception of the perfect self. What a person *should* be was friendly, happy, popular, confident, comfortable with itself and with strangers. What a person *shouldn't* be was me. I'd compared my real self to the fantasy model and decided I was getting it wrong. I'd tried to change what couldn't be changed and berated myself for failing. In doing this, I was hardly alone.[9]

In this account, the pursuit of self-esteem presents itself as a debilitating, incapacitating endeavor potentially harmful to emotional wellbeing. Perhaps if we freed ourselves from this "myth of the perfect life" and the endless pursuit of happiness, we would feel less stressed and, ironically, much happier.[10] Oliver Burkeman astutely argues that, "despite our fixation with happiness, we do not seem very happy. Modern life seems to have done little to lift our collective mood. It seems that increased income and economic growth has not resulted in happier societies. Nor have self-help books, so it seems."[11] At worst, pursuing happiness limits our emotional experience by dissuading us from exploring and experiencing our full spectrum of emotions.[12] In this chapter, readers will explore the origins of these developments.

From self-conceit to self-esteem

Victorians' ideas about self-esteem were concerned more with its fragile links with self-conceit (or excessive, unwarranted pride) and resultant impacts upon character. Unlike today, emotional health and wellbeing were not prime concerns. Excessive vanity was considered a "darker passion." In 1822, one author coupled too much self-esteem with "envy, malice and all un-charitableness," adding that it was "the deadliest foe of truth and reputation," a trait that "tears the laurel from the hero's brow and the wreath from the poet's tombs."[13] Nineteenth-century contemporaries undoubtedly saw benefits in having a certain amount of self-regard. As the *Sixpenny Magazine* bluntly stated in 1865: "A person who is not blessed with a fair opinion of his own abilities and acquirements will in nine cases out of ten turn out in the end to be a nobody." Nonetheless, unwarranted self-esteem turned people pompous and disagreeable. It was far better to seem modest than overconfident.[14]

Unlike in the twentieth century, when self-esteem became rethought as a limitless resource to be constantly tapped into, nineteenth-century physicians and psychiatrists sought to regulate, not optimize, the "passions" and "sentiments."[15] Phrenology was a nineteenth-century science that involved studying the lumps and bumps of skulls in the (erroneous) belief that these provided insight into mental faculties and personality traits. American phrenologist Silas Jones wrote in 1836:

> Self-esteem, more perhaps than any other feeling, should be a well-regulated and enlightened principle ... self-ignorance is the cause of the odious manifestation of self-esteem. Let it be active, in harmony with all the other sentiments, and it confers great dignity upon character. When in adversity, we need its utmost power, to balance those other feelings which have become morbidly excited. When in prosperity, we should watch over it constantly or it will lead us into a snare.[16]

Jones favored a balanced, regulated character in which self-esteem was present but not omnipresent, constrained but not overpowering and self-defining.

The foolishly prideful were disdained. Writing in 1874 about social reformer Robert Dale Owen's autobiography, William Brighty Rands (using his pseudonym Henry Holbeach) produced this scathing account of unjustifiable and delusional self-confidence:

> Self-esteem is nearly always large in benevolent people. Philanthropists as a class have it strongly marked. Robert Owen had no end of it! ... The organ

of self-esteem is, as a rule, well-marked in foresters, political teetotalers, and, indeed, all lovers of organizations which have very exclusive pretensions or which seek to impose fresh restrictions upon others ... and you will be more fortunate than I have ever happened to be, if you succeed in convincing such persons that they deserve to be horsewhipped.[17]

Rands's reference to "the organ of self-esteem" also drew from phrenology. Phrenologists believed the brain to be the organ of the mind (a relatively novel idea at the time) with specific areas having localized functions.[18] Franz Josef Gall identified an "organ of self-esteem" in the upper back of the head. Upon discovering a large prominence in that area, phrenologists concluded that the person under their examination had much self-pride, which would be a problem only if it manifested in arrogance, selfishness, and inflated self-importance. A smaller prominence indicated modesty, humility, and lower self-opinion.[19]

Phrenologists could easily spot the physical signs of excessive pride. Hewett C. Watson wrote that "those so endowed have generally an upright gait, carry their heads high, and have altogether an air of consequence about them. They are apt to speak in a pompous measured style, as if every word they uttered was highly oracular. They are great egotists, indulging largely in the use of the pronoun 'I' and talking constantly of their own affairs." Excessive self-esteem was generally associated with men.[20] Notably, phrenologists tended not to dwell too much on low self-esteem and its symptomatic relation to mental or physical disorder. Their concern lay with the pitfalls of character traits, not emotional sickness. A truly medicalized vocabulary of self-esteem was yet to develop.

During the nineteenth century, and across the West, emotions emerged as a distinct psychological category of analysis, replacing older concepts such as "passions," "sentiments," and "affections."[21] In this context, self-esteem was rethought in recognizably modern psychological terms as a more desirable personality trait linked to emotional wellbeing. (Of course, I do not suggest that society ever came to view arrogance positively.) Self-esteem provided an alternative interest to psychoanalytic understandings of narcissism and excessive (but inauthentic) self-love, much discussed early in the twentieth century, referring to healthier levels of self-love somewhere below the narcissistic scale less associated with arrogance and selfishness. Self-love was renovated into something more positive as long as it was properly regulated.[22] By the 1960s, self-esteem was central to the emerging "positive psychologies" (as I shall explore below).

To understand what ultimately made these interpretations of self-esteem so appealing, it is useful to reflect on what it was they replaced: the largely negative, gloomy ideas of human nature that predominated in the late nineteenth and early twentieth centuries. From the 1850s, Charles Darwin's evolution theory rooted human behavior firmly in biology. Subsequently, degeneration theory and Social Darwinism (both of which underpinned eugenics) maintained that human defects and problems (e.g. insanity, crime, alcoholism) were hereditary, organic (or physical) in origin, and largely irreversible.[23] This left little hope for those deemed "unfit," a group seemingly locked in a Darwinian struggle for existence against the "fitter" members of the human species. Some eugenicists thought it best for them to naturally die out. Others feared that the "unfit" might eventually out-reproduce the fit, who would then be losers in Darwin's battle of the species. To prevent this social calamity, eugenicists advocated, with varying levels of enthusiasm, enforced sterilization, birth control, and, at their most extreme, extermination.[24]

Such ideas promoted a distressing, distrustful outlook on humankind and its nature. If the "unfit" were a lost cause, their problems biologically ingrained and irreversible, then the stream of "bad life" must be stopped by discouraging, or making impossible, reproduction among some social groups. This pessimistic intervention took place firmly at the biological level. This was a radically different view of humankind than the one espoused later by the self-esteem movement, which saw human potential as boundless and urged endless self-improvement. Self-esteem theory promised hope for all; Social Darwinism, at its worst, saw hope only in ending or preventing life.

Freudian theory had a similarly pessimistic streak. Whereas eugenics found human deficiencies to be inborn, Freud emphasized life developments and held out some hope of possibilities of positive individual change. Nonetheless, many considered his theories too melancholic. To oversimplify, Freud's basic premise was that humans are driven by repressed experiences situated in their unconsciousness, typically linked to infant sexual experiences, development, or feelings. Freud developed psychotherapeutic methods to help patients manage, or better understand, their emotional behavior. However, he presented a world of inner psychological turmoil in which patients were held back by their repressed memories. Intervention was achievable only through psychotherapy, offering little scope for self-directed improvement. Psychoanalysts needed to be present to unravel their patient's life histories.[25] After witnessing the massive death rates of World War I, Freud redefined civilization as a space of conflict, reflecting a broader interwar cultural pessimism and antimodernism. In Freud's view, humankind's efforts to master instincts such as aggression and self-destruction were not altogether futile, but would always be partial.[26] Freud saw unhappiness as a normal, natural condition, and even suspected that we could never attain full happiness.[27]

Although Freud never developed a complete theory of self-esteem, he discussed how, as we develop, we adopt negative and positive attitudes toward ourselves. We can attain more positive views of ourselves by nurturing "positive self-regard" stemming from the ego and the superego. Even despite its generally gloomy nature, Freud's work helped provide some basis for the idea that feeling good about oneself might help defeat guilt and self-blame and promote self-worth. Nonetheless, his interest was in working through problems caused by past experiences to reduce misery in the present.[28] The negative aspects of Freud's work also failed to appeal to the general public in quite the same way that later positive psychology approaches did.

From character trait to emotional sickness

Some neo-eugenicists still see hope, controversially, in biological regulation and intervention.[29] Freud's basic concepts and psychological language are still used today, from unconscious thoughts and ego to the meaning of dreams.[30] However, as self-esteem theory developed, it came to support more optimistic understandings of human inner life that saw infinite potential for personal and emotional growth and, perhaps most importantly, for happiness. Interest shifted from moral "character" toward "personality," driven largely by the burgeoning field of psychology.[31] Nonetheless, although the concept of self-esteem was redefined, its underlying problems stayed the same: individualism, victim-blaming, lack of social context, constant self-regulation, and the pathologization of emotions other than happiness.

Typically, psychologists trace the origin of medicalized understandings of the self-esteem concept back to William James, an American philosopher and psychologist usually described as the founder of functional psychology. Shedding psychology's roots in philosophy, James developed a practical, therapeutic discipline. Nowadays, psychologists regard James's *Reflections on the Principles of Psychology* (1890) as an intellectual milestone.[32] James established a practical therapeutic program based on the scientific study of consciousness, borrowing techniques from European laboratory science to give psychology clinical utility and a scientific sheen. In relation to self-esteem, James was interested in the feelings that people generate when evaluating their achievements. The more successful someone is, the better they feel about themselves. For James, self-esteem was determined by the ratio between achievements and expectations, ambitions and realities, hopes and actualities. The closer their alignment, the better people felt. James encouraged his readers and patients not to "boost" self-esteem but instead to line up achievements and aspirations.[33] To accrue self-esteem, people could meet their expectations or lower them to realistic levels.[34] Although influential, James's popularity waned as Freud's rose.

Particularly from the 1930s, social psychology gained influence, a fundamentally political, as well as therapeutic, endeavor. Western social psychology largely mirrored the values of democratic, as opposed to authoritarian, governance. Psychological ideas reflected, but also contributed to, liberal agendas of ruling and governing citizens without impinging on their freedom and liberty, an approach that distinguished Western democratic culture from authoritarianism. American social psychologists, largely in response to their perceptions of Communist Russia, Fascist Italy, and Nazi Germany, developed theories and practices that encapsulated Western democracy and individualism. To help governments intervene indirectly in the everyday lives of citizens, they developed a rational knowledge of how humans operated in their lives: their motivations and interactions. They supported liberal governance by educating, not forcefully directing, citizens in ways that supported personal freedom.[35]

In the 1930s, American philosopher and sociologist George Herbert Mead built on James's concept of the "social self." Mead was interested in how people integrate into social groups. He proposed that, while integrating, they internalize the ideas and attributes of key figures in their lives, adopt these attitudes in their self-image, and then unknowingly express and experience other peoples' opinions as their own. In this interpretation, self-image reflected how "significant others" saw us. Self-esteem levels were contingent upon whether other people admired and praised us, or rejected and ignored us. Self-esteem was a mirror image of other people's perceptions. Those with high self-esteem had probably been treated kindly and respectfully; those who felt lowly had been treated cruelly and harshly.[36] Alongside sociologist Charles H. Cooley (who developed ideas of the "looking-glass self"), Mead supplied much of the theoretical framework that underpinned self-esteem research.[37] James and Mead approached self-esteem differently. James measured the gap between aspirations and reality; Mead concerned himself with the influence of social interactions. However, both framed self-esteem as a fluid emotional state, one that could be modified by either meeting personal aspirations or improving social interactions. Self-esteem levels were not set in stone; instead, they changed and fluctuated over time.

Nonetheless, until the 1960s, psy-experts focused largely on pathologies and sickness. Early twentieth-century psychology and sociology were rife with concerns about anomie, alienation, and estrangement, and, importantly for our purposes, inferiority complexes. Low self-esteem was now becoming medicalized, but also pathologized. Austrian psychotherapist Alfred Adler first developed the idea of an "inferiority complex." Born in 1870 in Rudolfsheim, a village on the outskirts in Vienna, Adler suffered various ailments in his childhood, including rickets, which prevented him from walking until the age of 4. He later decided to train as a physician, specializing initially as an eye doctor. In 1902, Freud invited him to join his informal discussion group, the Wednesday Society. Adler remained in Freud's close-knit immediate circle until 1911 when he disengaged due to personal and professional differences. Until his death in 1937, Adler enjoyed international success and celebrity.[38]

Adler's primary point of departure from Freud was his insistence that the social realm was as important to personality formation as the inner realm (although Adler's emphasis was on families and interpersonal relations). Like Freud, Adler investigated the power dynamics between the id and the super-ego, and childhood experiences. He theorized that the all-controlling power of adults caused a natural inferiority complex in infants. However, this served a useful purpose: it encouraged healthy emotional development in infants by fostering a strong determination to join the adult world. Feelings of inferiority motivated infants; problems arose only if parents failed to nurture this emerging self-confidence, haplessly inculcating a deeply ingrained feeling of inferiority that affected adult life. Such people could adopt a false front of self-confidence, but, by doing so, they risked developing a superiority complex, in some ways similar to narcissism. This was a fragile self-esteem, a façade that masked lowly inner feelings.[39]

Despite foregrounding self-esteem's negatives, Adler still hoped that a positive, scientific psychology could help reform society for the better. Indeed, some scholars see Adler as a precursor of the positive psychology movement that later blossomed.[40] Adler thought that psychologists could help people better understand the internal problems that unknowingly frustrated them only if they moved on from dealing with sexual trauma alone. Adler also challenged ideas about the futility of addressing hereditary problems. He argued that what humans inherited was less important than how they worked with that inheritance. Even if someone inherited negative defects, psychology could still help.[41] This opened up new hopes for self-improvement not immediately obvious in Freudian theory or eugenics programs. Adler's inferiority complex concept was received particularly warmly. Unlike Freudian psychoanalysis, which many regarded as morbidly depressing and overly sexual, the inferiority complex idea seemed more accessible.[42]

Harry Stack Sullivan was another neo-Freudian interested in the social origins of personality and interpersonal relations. Influenced by Adler, Sullivan believed that personalities could only truly be understood in relation to social interactions.[43] He warned that people constantly guard themselves against losing self-esteem to avoid anxiety, defined as an interpersonal phenomenon that arises when people are, or are expected to be, rejected or demeaned by others. Although only barely touching on the hows and whys of maintaining self-esteem, Sullivan highlighted the primary importance of parents and siblings.[44] Like Mead and Adler, he saw self-esteem as variable in particular situations and periods of life.

And, like Adler, he stressed the harmful consequences of low self-esteem, focusing on anxiety, not inferiority. Sullivan also situated the causes of low self-esteem in interpersonal and familial environments. Ultimately, his work represents a path not taken in American psychotherapy; his colleagues considered him intellectually difficult and politically opinionated. Unlike them, Sullivan was unafraid to address controversial issues such as racism, economic injustice, and nuclear war. He was also a critic of individualism.[45]

Finally, German psychoanalyst Karen Horney further helped develop and popularize ideas about self-esteem. Initially working within a Freudian framework, Horney published a series of essays on feminine psychology in the 1920s and 1930s. In the late 1930s, she published two books that rejected some of Freud's most basic premises. Her emphasis was on culture and interpersonal relationships, and particularly on how people developed defense mechanisms to cope with the anxiety and frustration of psychological needs. For Horney, there existed no universal human nature or neurosis: these were culturally contingent. The origin and structure of neurosis rested in adverse environmental conditions, usually family environments, which created anxiety and encouraged infants and children to develop defense mechanisms. In Horney's view, sexual difficulties were the *result* of personality problems, not their *cause*. She advocated exploring current defense mechanisms and inner conflicts, not those hidden in an infant's past, deviating, like others, from basic Freudian approaches. This emphasis on the present also clashed with orthodox psychoanalysts, and Horney ended up resigning from the New York Psychoanalytic Institute.

In her later works, *Our Inner Conflicts* (1945) and *Neurosis and Human Growth* (1950), Horney developed ideas of a "pride system" whereby people compensated for feelings of worthlessness and inadequacy by developing idealized images of themselves. At worst, this created neurotic pride and inner self-hate. The impossibility of attaining this idealized image intensified self-hate, creating what Horney termed a "vicious circle." While different in principle from Adler and Sullivan, Horney's ideas stressed the importance of developing "real" self-esteem, situated the cause of emotional disorder in individuals and families, and encouraged people to realize their human potential through emotional self-work.[46] Horney was well-known in her lifetime, and enjoyed a later popularity surge from 1967 when her essay collection, *Feminine Psychology*, was published.[47]

Combined, these theoretical developments brought new discourses, practices, and techniques for monitoring, regulating, and adjusting self-esteem levels. Discussion of self-esteem among psychologists turned toward emotional sickness and dysfunction, displacing the Victorian emphasis on character and conduct. New concerns arose about inferiority complexes, defense mechanisms, anxieties, and neuroses, although these ideas were inherently individualistic and divorced from sociopolitical context. Solutions to life's problems involved adjusting aspects of individual or familial life, without critically thinking about socioeconomic environments and structures. Not that this stopped self-esteem from becoming central to psychological thought. Nor did it discourage the American public from developing faith in the concept as a key to happiness, prosperity, and emotional wellbeing. On the contrary, self-esteem fitted well with American individualistic ideals and, later, those of neoliberalism. The idea took off in the individualistic, consumer culture that followed immediately on from World War II, and blended well with the capitalist culture that accelerated from around the 1970s. All of this formed part of a broader drive in which Westerners began to understand and relate to themselves (and others) as psychological beings.[48]

Self-esteem and popular culture

After World War II, American psychology developed massively. The American Psychological Association (APA) made clinical training mandatory on university-based psychology courses. The Rockefeller Foundation endowed university departments and research centers with financial support. In 1946, the US National Mental Health Act allocated increased funding for psychological research and education and also created the National Institute of Mental Health (NIMH), a new center for psychological research. Psychologists moved away from examining only "abnormal" patients to exploring everyone's emotional wellbeing, even the healthy and the "normal."[49] The immediate postwar years were a time when Americans truly absorbed psychological ways of thinking about themselves. Ego, inferiority complex, self-esteem: these concepts all became household terms.[50] These developments dovetailed with a burgeoning public interest in self-help. In the context of immigration, consumerism, urbanization, decline of extended family ties, and intensification of the "society of strangers," psychologists as well as self-help gurus now filled a void once occupied by closer family and community connections.[51]

Did all of this have much impact on American self-esteem levels, for better or for worse? This is a difficult question to answer. A few historians argue that a discernible change might well have occurred. The self-confidence enjoyed by the baby boom generation may have arisen from the child-centric outlook of white middle-class families that was popular from the 1950s, resulting in smaller families, greater prosperity, and increased attention lavished on children.[52] However, others insist that widespread low self-esteem increased, largely due to heightened isolation, emotional distress, and excessive self-reflection.[53] What is clear is that personalities replete with self-esteem became upheld as a cultural ideal, norm, and aspiration. Presumably, this helped some people attain a positive self-image. Perhaps some people even felt happy. However, those who felt persistently underconfident risked being made to feel "abnormal," even alienated from a culture that venerated personality types vastly different from their own.

By the 1950s, Americans had many reference points for improving their self-confidence. Men's magazines such as *Cavalcade* (the self-styled "know yourself magazine") ran articles with titles such as "Beat Your Inferiority Complex." These urged men to realize that, despite their inner doubts, women were attracted to them, they could play golf as well as anyone else, they were as intelligent as the average university graduate, and it's okay to be short. All that was needed to feel good about oneself was a focus on strengths, not weaknesses.[54]

Meanwhile, *Cosmopolitan*, aimed at women, ran stories on the emotional wellbeing benefits of beautifying. "When a woman looks her best and knows it," claimed one 1956 article, "the realization infuses a feeling of self-assurance that communicates itself automatically to the people she meets and makes them more favorably impressed with her. This in turn increases her own feeling of self-esteem." Evidence was found in the tale of a beauty consultant having visited New York's Fountain House, an institution that reintegrated former asylum patients into the community, to deliver a lecture and beauty-care demonstration.

Twenty women gathered around her who initially seemed depressed and uneasy until they were told how to improve their appearance with beauty products. Their new-found confidence made them feel just as efficient and useful as anyone else, reported *Cosmopolitan*.[55] Of course, such accounts reveal the gendering of self-esteem. Men gained confidence from social achievements, women from physical appearance. However, we can see that the new dream of self-esteem, confidence, and happiness was now being targeted equally, albeit differently, at male and female readers.

Contemporaneously, a new genre of self-help literature emerged that focused initially on "positive thinking." Until the 1950s, positive thinking literature mostly encouraged readers in terms of economic gain. Dale Carnegie's *How to Win Friends and Influence People* (1936) was among these. Carnegie had no quibbles about faking self-confidence to accrue wealth.[56] Similarly, Napoleon Hill's *Think and Grow Rich* (1937) enticed readers to think positively to accumulate wealth.[57] These publications were more about self-advancement than self-growth, but from the 1950s, self-help literature drew more extensively from psychological theory, and promoted positive thinking for its emotional, as well as material, benefits. Melvin Powers's *Positive Thinking: The Technique for Achieving Self-Confidence and Success* (1955) drew directly from psychoanalytic concepts of the conscious and subconscious for the purpose of making "favorable personality changes" to achieve "success, self-mastery, self-confidence and personal happiness." Powers advised that "if we saturate our minds with the idea of success, success will finally come to us. For one is a product of one's thinking." His practical techniques included spending up to ten minutes each day painting mental pictures of the things that one desires in life, then imagining oneself in those pictures. Testifying to the individualizing tendencies in postwar self-help, Powers also advised: "Forget about the Cold War and pessimistic world affairs. These are distracting. Instead, find a ray of hope and be uplifted in the thought of a right future. Focus on self-confidence, not world politics."[58]

Some self-help books combined therapeutic and religious approaches. Norman Vincent Peale was a mainstream Protestant minister initially attracted to the early twentieth-century New Thought movement. Sales of Peale's books and printed sermons rose significantly after World War II. *The Power of Positive Thinking*, first published in 1952, is still in circulation today. His approach has been described as "applied Christianity," a spiritual method that overlapped with emerging therapeutic sensibilities in American culture.[59] Although Peale aimed his work at everyone, it especially targeted the lonely 1950s salesman thought to be in urgent need of a self-confidence lift.[60] Peale insisted that our lives are shaped largely by other people's attitudes toward us, and that happiness is key to prosperity. Essentially, he promoted mental re-programming and conditioning. If negative thoughts replaced positive ones, inner happiness and success would naturally ensue. To reinforce his message, Peale discussed his own inferiority complex as a child arising from being too thin and desiring to be muscular. As an adult religious minister, he forced himself to overcome a fear of public speaking, finding solutions in both inner mental processes and faith in God.[61] Although he stressed the importance of the unconscious mind, ultimately only faith in God could unleash full potential. *The Power of Positive Thinking* consists of a list of exercises intended to encourage readers to think positively.

Aware of his potential to depoliticize personal experience, Peale took care to note that he didn't intend to ignore or trivialize the world's hardships. Nonetheless, he didn't want global problems to dominate his readers' thoughts. By engaging with positive thinking, readers could overcome obstacles by transforming into popular, esteemed, well-liked individuals with a new-found sense of wellbeing, health, usefulness, and influence. "Believe in yourself!" instructed Peale in his opening pages. "Have faith in your abilities! Without a humble, but reasonable confidence in your own powers you cannot be successful or happy. But with sound self-confidence you can succeed." "It is appalling to realize the number of pathetic people who are hampered by and made miserable by the malady popularly called the inferiority complex," Peale added, "but you need not suffer from this trouble. When proper steps are taken, it can be overcome. You can develop creative faith in yourself. Faith that is justified."[62]

Despite the ready availability of self-help advice, not everyone began to feel confident. A burgeoning culture that privileged self-confidence potentially alienated anyone whose emotional realities failed to live up to happiness standards. By underscoring self-esteem's importance, American culture regularly made those lacking positive feelings feel at odds with society. We may see here some aspects of the "looping effect" at work. In the 1940s and 1950s, Hollywood star Rita Hayworth publicly assessed her string of failed marriages as a consequence of an inferiority complex.[63] Movie director Vincent Sherman later stated: "Rita was very wonderful and very talented but the saddest girl I've ever known. She had been used by every man who ever worked with her. They all went to bed with her and used her. She was very shy, uneducated, very insecure and had a terrible inferiority complex about herself."[64] After a 1950 suicide attempt, Judy Garland confessed to suffering from "a mild sort of inferiority complex."[65] Marilyn Monroe had many encounters with psychoanalysts. In 1956, she sought the help of Anna Freud, psychotherapist and Sigmund's daughter. That same year, Monroe's director, Joshua Logan, suggested: "She has immense subtlety, but she is a frightened girl, terrified of the whole film-making process and self-critical to the point of an inferiority complex."[66]

Men were not immune either. James Dean represented a new type of masculinity. His on-screen characters cried, struggled, screamed with frustration, and expressed a need for closer relations with their fathers. Dean was an emotional man, skillfully blending machismo and sensitivity. He was self-assured, while also vulnerable and tormented.[67] The 1957 documentary, *The James Dean Story*, revealed that Dean carried around with him a short pamphlet entitled "The Complete Man's Three Needs" to improve his confidence. The needs were (1) love and security, (2) creative expression, and (3) recognition and self-esteem.[68] In early interviews, Elvis Presley stared at his feet and froze with nervousness. However, after releasing his first record, he developed his self-confident trademark swagger. Reportedly, Elvis memorized entire passages from films starring James Dean and Marlon Brando, both of whom portrayed characters who, on the surface, exuded self-confidence, charm, and charisma.[69] One of Presley's film directors, Gene Nelson, later commented that Elvis "suffered an acute lack of self-esteem as a human being … he felt that he was uneducated and had nothing to contribute to a conversation."[70]

Recent reassessments of Brando similarly emphasize how he felt haunted by memories of an alcoholic mother and abusive father, and how he battled low self-esteem and loneliness. In his own words: "I had a great feeling of inadequacy, that I didn't know enough, that I didn't have enough education. I felt dumb … inferiority has been very close all my life."[71] These were icons who had successfully worked on their self-esteem, at least on the surface. These stars drew from an increasingly influential psychological language to articulate their inner feelings. But could it have been that cultural emphases on feeling positive and confident had worsened their self-perceptions in the first place?

Pathological personalities

It was undoubtedly beneficial that psychologists now recognized low self-esteem as a significant psychological condition, one with etiological connections to a range of emotional problems, for this opened up possibilities for therapeutic intervention that had been unavailable in the Victorian emphasis on character. Psychology offered an appealing vocabulary for celebrities and self-help readers alike to draw from.[72] The diverse, but interconnected, theories outlined above situated the root cause of personal and social problems in malleable day-to-day activities and interpersonal interactions, offering hope in improving inner feelings. Nonetheless, they also urged Americans to internalize life's difficulties. If improvement failed to materialize, the individual must still be at fault. Self-help encouraged those suffering low self-esteem to *adjust* to their surroundings without thinking much about society's overarching structural problems. In his influential 1955 book, *The Sane Society*, psychoanalyst Erich Fromm influentially arraigned Western culture (and the psychology profession) for encouraging individuals to adjust to a society that was itself maladjusted. Fromm considered capitalism to be dehumanizing and alienating.[73] Could it be that capitalist society was sick and insane, and not the individuals within it? Was society pathological, rather than people? Influential feminist activist Betty Friedan voiced similar concerns about efforts made by psy-experts to "cure" patients by encouraging their assimilation into a patriarchal culture.[74]

When viewed from this perspective, the transformation of low self-esteem into a quasi-medical category appears more problematic, not because a meaningful emotional state had been identified, but because overpromoting self-esteem could have injurious consequences. Psychologists began to refer routinely to the self-esteem concept when determining whether someone had a healthy or a sick personality, delineating sharp differences between "normal" and "pathological" feelings.[75] From the 1960s, self-esteem levels became quantifiable once prominent social psychologist Morris Rosenberg developed a ten-point self-completion tool. The fragile boundaries between "normality" and "abnormality" could now be measured, which helped embed self-esteem concept in clinical practice.[76] However, it was imperative to be situated at the higher, positive end of Rosenberg's scale. Not that self-esteem had any finite boundaries; self-esteem was limitless, the potential for self-improvement ceaseless, but so too was decline into negative psychopathology.

New personality types were sketched that offered models for Americans to aspire to or altogether avoid. Adler helped normalize self-confident personalities by pointing out the dangers of failing to nurture self-esteem:

> Only such persons as are courageous, self-confident and at home in the world can benefit both by the difficulties and by the advantages of life. They are never afraid. They know that there are difficulties, but they also know they can overcome them. They are prepared for all the problems of life, which are invariably social problems.[77]

For Adler, those blessed with high self-esteem were "useful" individuals with a strong "awareness of social interest." By contrast, those lacking self-esteem were "oriented toward the useless side of life." These included problem children, criminals, the mentally ill, and drunkards.[78] Adler constructed a distinct image of the psychologically damaged, under-confident person frustrated by a constantly niggling inferiority while traversing their daily life. He created the antithesis of the self-confident person, a vulnerable individual whose inner life was abnormal and in need of expert intervention. Admittedly, Adler chose not to leave his introverted patients doomed to lifelong misery. He thought their situations might improve once an inferiority complex had been identified. Ultimately though, Adler was dismissive and judgmental toward those who failed, or even refused, to work on their self-esteem. He also promoted a sense that society was not at fault for manifestations of psychological disorder; rather, individuals were. And it was now not only the criminal and deviant who needed to keep watch on their self-esteem levels, for everyone might be emotionally sick without even realizing. For Adler:

> Complex is not the correct word for this feeling of inferiority that permeates the whole personality. It is more than a complex. It is almost a disease whose ravages vary under different circumstances. Thus, we sometimes do not notice the feeling of inferiority when a person is on his job because he feels sure of his work. On the other hand, he may not be sure of himself in society or in his relations with the opposite sex, and in this way, we are able to discover his true psychological situation.[79]

Here, Adler framed everyone as potentially harboring feelings of inferiority, reflecting psychology's burgeoning extension from the clinic to society at large. As he explained:

> [Everyone] has a feeling of inferiority, but the feeling of inferiority is not a disease. It is rather a stimulant to healthy normal striving and development. It becomes a pathological condition only when the sense of inadequacy overwhelms the individual and, so far from stimulating him to useful activity, makes him depressed and incapable of development.[80]

If self-esteem was not worked on, people would continue to lack the inner resources needed to tolerate modern life. In other words, everyone had to monitor their self-esteem to adapt to their surroundings, without reflecting much upon the problematic nature of those surroundings.

In light of all of this, acquiring self-esteem needed to be a lifelong project. Positive self-esteem needed to be genuine, not a false front. Adler situated those who exuded false self-confidence in his "useless side of life"; these individuals were not socially adjusted, regardless of outward appearances.[81] Horney similarly believed that self-esteem needed to be sincere. It was not enough to artificially exude a self-confident aura. Anyone who attempted to mold themselves into beings of absolute perfection risked becoming alienated from their real selves. "For nothing short of godlike perfection," wrote Horney, "can fulfil his idealized image of himself and satisfy his pride in the exalted attributes which (so he feels) he has, could have or should have." It was only by developing authentic *depth* of feeling, thoughts, and interests that Americans could use their inner resources to reach emotional maturity.[82] Hayworth, Garland, Monroe, Dean, Brando, and Presley would all have known full well that the impression of confident sexuality expressed through their acting was not paralleled in their inner worlds. It was certainly possible to pull off a solid performance of having lots of self-esteem, but this was a shaky foundation for authentic psychological health.

Nowadays, ethicists encapsulate this using the concept of "bio-morality," which describes how Western society places moral demands on us to be happy and emotionally healthy. Happiness, positive thinking, high self-esteem: none of these is optional. They're obligatory.[83] While we're free to refrain from pursuing them, the morally persuasive messages inherent in self-help advice dissuades us from doing so. We can observe some of the origins of this bio-morality in Adler's writing. Appearing happy and confident is not good enough, for positive feelings should be firmly integrated into one's persona. Fake, artificial happiness simply will not do. And if we can all attain happiness through simple measures such as positive thinking, there's little excuse not to do so. However, there's a harsh insistence on personal responsibility in much self-help advice. If we don't feel confident enough, we aren't trying hard enough. In some ways, positive thinking can be compared to Calvinism with its harsh judgementalism and relentless self-monitoring. In the twentieth century, positive thinking ceased being a cure for the ill, anxious, and psychosomatically distressed; it became an obligation imposed upon us.[84] It contained an underlying imperative for constant change. Something always needed fixing or improving. There was no time to relax or feel fully happy with ourselves. But the end goal of optimal improvement always remained elusive.[85]

This approach was myopic and introspective. Particularly after the war, psychologists' rarely looked much beyond the individual or the family. Horney stressed that only children raised in emotionally healthy familial environments developed positive attributes and potentials. This milieu was marked by warmth, freedom of expression, encouragement, guidance, and rational discipline. Children raised in neurotic atmospheres became insecure, distrustful, and lonely ("basic anxiety"), and developed defense mechanisms to guard against their flailing self-esteem.[86] Here again, the problem of low self-esteem was situated within the immediate (presumably nuclear) family, obscuring the social-cultural contexts in which familial experienced and lived through their emotions.

The problem, then, with acknowledging low self-esteem as a therapeutic one rested more in how the concept was used. Low self-esteem wasn't seen as a sickness in itself but was now considered abnormal, antisocial, and indicative of deeper psychological distress. Psychologists cast those lacking self-esteem as being not "useful," "adequate," or "socially adjusted." They undoubtedly set out with good intentions when idealizing self-confidence, but by raising the social capital of the self-confident, they alienated the introverted and the underconfident. As poor self-esteem affected everyone around us, failure to cultivate self-esteem was now viewed suspiciously, even as socially irresponsible, for, according to this perspective, low self-esteem was the *cause* of social problems, not one of their symptoms. Accordingly, working on one's self-esteem became socially imperative.

Self-esteem and the counterculture

The 1960s counterculture was a curious mixture of sociopolitical activism and personal introspection. In a 1969 edition of the radical newspaper *Los Angeles Free Press* could be found advertisements for a weekly self-esteem and self-acceptance workshop, alongside adverts for workshops on women's liberation, poetry, Hindu teachings, native Indian integration and s.e.l.f., a nude, nonverbal workshop.[87] The Beatles encouraged us to explore our inner life. "Try to realize it's all within yourself," sang George Harrison in "Within You Without You."[88] During 1968, many hippies felt just as enthusiastic about inner LSD trips and meditation as they did about staging violent uprisings.[89] The diverse forms of therapy and introspective techniques aimed to produce healthier, more fulfilled individuals who would reach a critical mass and take society in a better direction. However, even the counterculture ultimately ended up being more individualistic than sociopolitical.

The counterculture was inspired by a range of ideas that emphasized internal life over postwar capitalism and materialism. Young people sought philosophies and worldviews that supported the search for personal development. Hinduism was prominent, while forms of Christianity that stressed adherence to preconceived, even authoritarian, ideas were refuted. Rejecting authority was a common thread in both the individual and the sociopolitical aspects of the counterculture. However, while advocating social transformation, the basic unit of change was still often the individual.[90] In his 1966 advice to "drop out," psychologist and LSD enthusiast Timothy Leary declared that political involvement was meaningless, for "any external or social action, unless it's based on external consciousness, is robot behavior."[91] Journalist and beat poet Lawrence Lipton, when discussing his enthusiastically anticipated erotic revolution, similarly encouraged readers to dis-educate and then re-educate themselves, to disentangle themselves from sociopolitical structures and instead build smaller enclaves or circles of friends who would sustain self-confidence through mutual support.[92] The Maharishi, who famously spent time with the Beatles in 1967–8, was skeptical about group political movements, regarding those unable to stand on their own as weak. Many hippies followed his example by retreating into meditation, cannabis, and LSD.[93] The new emphasis was on personal, not traditional, party politics, with a healthy supply of radical activism where necessary.

The counterculture fundamentally rethought what politics was about. In part, it became a struggle for the liberation of feelings in a society dominated by rationality.[94] Leftwing radicals often spoke of people feeling damaged in modern life and by postwar capitalism's production of fear, loneliness, and social isolation. Activists considered capitalism in emotional terms, imagining their politics as a struggle against rationalities that created negative emotions. Techniques involving self-esteem offered to liberate citizens from the emotionally miserable aspects of capitalism. Citizens were to free themselves from, not simply adjust to, postwar society.[95] In this respect, self-esteem strategies were fundamentally liberating.

Despite their anti-expert stance, 1960s social movements drew freely from psychological ideas. There was a considerable growth in ideas of personal development, as promoted by the Human Potential Movement, so named by novelist Aldous Huxley.[96] Like other humanistic psychologists, this group self-consciously developed new forms of psychology. Questioning Freud's gloomy obsession with sexual trauma, they instead delighted in humankind's potential for self-realization. They queried psychology's obsession with negative psychological states. More hopeful, they replaced discussion of emotional weaknesses and deficits with capacities and potentialities.[97]

Carl Rogers was a key advocate of humanistic, or "client-centered," psychological approaches.[98] Rogers was born in Chicago in 1902.

Although raised in a deeply religious household, he became an atheist, but retained an interest in the mystical, spiritual, and transcendental. From the 1930s, he developed a curiosity in Freudian psychotherapy while working in child study. In 1940, he became professor of clinical psychology at Ohio State University. Throughout the 1950s, he published his key works on client-centered therapy, before spearheading humanistic psychology.[99] Rogers represented society not as something to which people needed to adjust, but instead as an omnipresent force from which the self must be liberated. Improving self-esteem might help free people from inhibitions as part of their ongoing liberation.[100]

Rogers developed the idea of the "actualizing tendency," which maintained that humans have underlying inclinations to work toward accomplishing their full potential and reaching the highest possible levels of "human-beingness." Such assertions contrasted starkly with older emphases on inferiority complexes, anxieties, and neuroses. In his influential *On Becoming a Person* (1961), Rogers developed ideas about the "fully functioning person," someone who is open to experience, able to listen to themselves and others, and who allows themselves to experience life and circumstances without feeling internally threatened. This functioning person lives in the present and trusts, rather than fears, personal experience. They rely on their own judgments and assessments, regardless of external criticism from others. Rogers proposed that humans develop self-image largely from early interactions. Self-image reflected the judgments, preferences, and shortcomings of familial and social settings. Permissive familial environments that support free expression of ideas without harsh evaluation encourage children to know and accept themselves. Conflicts could be averted if family members were to accept the views and values of children.[101]

Abraham H. Maslow was another humanistic psychologist, who depicted Freud as the "sick half of psychology." Like Rogers, Maslow chose to focus on the "healthy half."[102] He was born in New York in 1908. During the 1920s, he trained as a psychologist. While working at Columbia University, he was mentored by Adler. From the 1940s, he helped develop humanistic psychology. The horrors of World War I inspired in Maslow a vision of peace and the need for positive understandings of humankind. Like Rogers, he regarded personal growth and life fulfillment as a basic human motive, despite having differing opinions on how to self-actualize.[103] Maslow is best remembered for his hierarchy of needs, the most basic of which are physical: food, water, warmth, safety. Once these have been satisfied, humans have two psychological needs: (1) esteem and (2) belongingness and love (intimate relationships, friends). The next stage is self-actualization (or achieving one's full potential). Maslow thought that everyone needs to satisfy their basic needs. Only once they have done so is time available to attend to non-essential psychological needs, and to self-develop and grow psychologically. Maslow admitted that our potential to self-actualize could be impeded if basic needs are left unfulfilled.[104] The basic idea of self-actualization arose in a postwar society of abundance. As hunger and shelter were generally less of a problem for the majority of Americans, time was freed up to work on "lesser needs." These ideas also fitted well into a socioeconomic climate that celebrated the pursuit of wealth.[105] (However, as subsequent chapters will demonstrate, minority groups fit uneasily into Maslow's model.)

The elevation of self-esteem to a position of centrality to human experience was mirrored in other literature. In his influential 1962 book *The Birth and Death of Meaning*, American anthropologist Ernest Becker devoted a chapter to "self-esteem – the dominant motive of man." Becker thought that Freud failed to satisfactorily explain human motives and considered Freudian theory "too grotesque and far-fetched in most cases." Becker insisted that "patients are still being rendered imbecilic by the psychoanalytic vocabulary of 'penis envy,' 'primary incest wishes,' 'the trauma of the primal scene,' and so on." Instead, Becker sided with Adler's interpretation of the urge toward self-esteem as a basic law of human life. As he explained, "the seemingly trite word 'self-esteem' is at the very core of human adaptation … the qualitative feeling of self-value is the basic predicate for human action, precisely because it epitomizes the whole development of the ego. This cannot be over-emphasized." Becker noted also that the human child is "the only animal in nature who vitally depends on a symbolic constitution of his worth." The entirety of adult life remains animated by the "artificial symbolism of his self-worth. Almost all his time is devoted to the protection, maintenance, and aggrandizement of the symbolic edifice of his self-esteem," passing through seeking approval from classmates and, later, in the workplace and marriage.[106] Similarly, in his 1969 book, *The Psychology of Self-Esteem*, hugely popular Canadian-American psychoanalyst Nathaniel Branden situated self-esteem at the very core of human motivation, behavior, and experience.[107]

Ideas about "growth" and "potential" departed significantly from earlier self-esteem theorists. Rather than sitting around dwelling on psychopathologies, those interested in human potential at the famed Esalen Institute in California were exposed to Gestalt Therapy, encounter groups, sensitivity groups, and Eastern techniques such as yoga, meditation, and tai-chi. Additionally, they were supplied with large quantities of LSD.[108] Nonetheless, psychologists such as Rogers and Maslow struggled to entirely abandon some of the core problems in self-esteem theory. By developing ideas that some of us think of as more "functional" than others, Rogers still implied that there existed many "dysfunctional" individuals, continuing the pathologization of certain personality types considered less ideal. Both Rogers and Maslow stressed the importance of family, just as Horney and others had, contradictorily reaffirming psychology's conservative tendencies. For such reasons, the Human Potential Movement's promotion of individualism in fact intensified the "isolation of the self," and the de-politicization and individualization of life's problems.[109]

These tensions between the individual and the social, the conservative and the radical, the political and the spiritual permeated the counterculture too. The political aspects of the 1960s counterculture gradually disintegrated, as the inherent difficulties involved in smashing the status quo became apparent. The same couldn't be said of the personal, spiritual aspects that evolved and survived, perhaps precisely because they posed less sociopolitical threat.[110] As Peter Doggett insists, ultimately "the personal became political, until the politics dissolved and the personal became all that existed."[111] As sociopolitical transformation remained frustratingly elusive, many counterculturalists retreated into the individual realm, resigning themselves to the fact that the status quo was not giving way any time soon.[112]

John Lennon's song "Revolution" seemed uncertain about whether a violent revolution was the best way forward or if everyone should "free their mind instead." Indeed, the lyrics sparked a protracted debate about Lennon's radicalism in the pages of *Black Dwarf*, a British underground newspaper to which Lennon contributed.[113] Lennon was at times overtly political, at other times spiritual and individualistic. In 1970, he sought the help of psychologist Arthur Janov, whose primal scream theory insisted that human neuroses could all be traced back to an initial moment of agony experienced by a child unable to satisfy its desires. If patients excavated decades of buried emotions, they could unleash the primal scream that would reconnect them with their own pain. The creative result was Lennon's self-titled debut solo album.[114] Nonetheless, his follow-up album, *Some Time in New York City*, marked a (less musically successful) return to the world of political activism, tackling sexism, the Northern Irish Troubles, prison riots, and activists Angela Davis and John Sinclair.[115]

However, as political activism receded, personal transformation for its own sake was all that remained. As hippy dreams faded away, self-esteem remained in its newly elevated position of centrality to human experience. Nonetheless, as this chapter has demonstrated, the seeds of many of the criticisms currently leveled at the self-help movement, and self-esteem specifically, had been planted. As the more political aspects of the counterculture became less prominent, self-help essentially became reduced to its more introspective elements, an emphasis on the personal with little reflection on the structural. As later chapters will demonstrate, this more individualized approach to self-esteem management became embedded from the 1970s into educational and self-help strategies which themselves largely supported capitalist, and also rightwing, political agendas. However, before learning how this happened, Chapters 2 and 3 will look more closely at self-esteem's radical potential for activist movements, and in particular at the positive uses found for the idea in African American and LGBT activism, both of which coincided with the 1960s counterculture movement. These will further support my view that the concept of self-esteem is multifaceted, and has served multiple purposes beyond just being a "tool of capitalism." It will also become clear that the models of psychological wellbeing discussed throughout this chapter largely failed to map on to the lives of marginalized groups whose identities and behaviors diverged from the new personality ideals glorified in postwar America, and which found themselves subject to types of social discrimination deeply embedded in America's socio-political structures.

TWO

Race

> For some individuals, being Black actually is advantageous as far as self-esteem is concerned ... Discrimination allows the Black man to explain his frustrations, failures and so on, rather than look to deficiencies in his own self.[1]

Liberal psychologist E. Earl Baughman, Jr. wrote these words in 1971, a time when African American activism had made considerable headway in challenging racial discrimination. This passage caught my attention (despite being written by a white psychologist), as it seemed to highlight some positive functions in the therapeutic turn for activists, the likes of which tend to be overlooked in sociological critiques of self-help that lambast the nature of modern capitalist life in more general terms. The previous chapter was equally guilty of laying out a generalized account of the development of self-esteem theory, linking this to therapeutic culture, consumerism, and the pursuit of happiness, while paying no attention to specific social groups. This chapter develops my views that we could move beyond thinking about therapeutic culture conspiratorially to reflect upon its complexity and multidimensionality, given that, despite its many problems, it has offered (and maybe still can do so) opportunities for socially marginalized groups to articulate emotional suffering and to support active campaigning against social injustice.[2] Nowhere was this clearer than in the racial politics of 1950s and 1960s America.

During the COVID pandemic, interest resurged in this history of self-care. The public began to remember that the civil rights and Black Power movements offered more than the oft-remembered marches and slogans. The Black Panther Party incorporated its own healthcare program into its struggle for social justice, viewing health as a basic human right.[3] While the party's interest in physical wellbeing is more often discussed, Black activists also absorbed psychological ideas. Feelings were crucially important to the Black Power movement and the construction of new Black identities. Black Power intentionally shifted the emotional mood from feelings of despair to empowerment.[4]

Self-esteem was arguably the most important emotion highlighted. As the idea of self-esteem took hold, and as Americans increasingly framed their sense of identity in relation to emotional wellbeing, some groups and individuals started to feel acutely aware that they weren't experiencing happiness in the ways encouraged by psychologists and positive thinkers. They weren't just failing to live up to the new American personality ideal; they felt actively excluded from it. The solution? Either continue to suffer emotionally or claim a right to emotional wellness. While white counterculture activists were busy navigating a self-indulgent path of inner exploration, Black activists used the concept of self-esteem in more urgent ways to address systemic racism. They pursued their endeavors against a white-dominated culture that labeled Blackness as inferior. Certainly, 1960s Black activists were far more able than counterculture activists to look beyond the confines of the mind and locate the source of emotional distress in political and socioeconomic structures working against their interests. Socially excluded groups were acutely cognizant of the structural causes of their negative feelings and how harmful emotional environments impact upon wellbeing.

Only a handful of historians have assessed the importance of self-esteem for 1960s Black activists.[5] Nowhere has the work of these activists been incorporated into a broader history of the self-esteem concept itself. This is not simply a Black versus white account. Liberal white psychologists such as Baughman examined the psychological nuances of Black problems. Nonetheless, Black communities lived through and endured feelings of low self-esteem, and ultimately turned to self-help for both day-to-day improvement and political empowerment.

To some extent, Black activists were successful. They genuinely fostered a sense of pride and self-esteem in their communities while persuasively linking Black emotional suffering to an oppressive political and socioeconomic climate. Their interpretation of therapy was transformative, not palliative. They intended it as a basis for revolutionizing, not adapting to, the "sick society."[6] Like many white critics of the day, African Americans identified capitalist society as sick, but they had their own specific problems to contend with, particularly racial inequalities.

Freedom might occur if Black people learnt to feel powerful, effectual, and in control of their own destinies. Self-confidence was key to a new, empowered African American identity. Self-esteem was not a personal means to an end; collectively, it might overturn injurious and discriminatory environments.

However, this approach had limitations. Despite the gains made, the gap between whites and Blacks persisted through the 1970s and 1980s, as did racial tensions. Only during the long recovery of the 1990s did significant economic improvements for Blacks become evident, and these, first and foremost, benefited the Black middle class. Then, in the early twenty-first century, the bankers caused the 2008 economic crisis, creating a devastating recession that reversed much of the progress made in Black lives.[7] The push for Black pride, esteem, and respect in the 1960s and early 1970s turned out to have been quite fleeting and momentary. Nonetheless, as Robin D.G. Kelley reminds us, we should not evaluate the work of social movements by whether they succeeded. We would be better off recognizing and exploring the meanings of the ideas, visions, and dreams that inspired Black people historically to struggle for change.[8]

"Negro inferiority"

To claim that a group is inferior is to exert power. To quote African American journalist and author Isabel Wilkerson: "Dehumanize the group, and you have completed the work of dehumanizing any single person within it."[9] Wilkerson describes race as a fluid, superficial idea; one that America's dominant white culture redefines periodically to suit changing socioeconomic and political circumstances and to sustain, according to Wilkerson, a caste system. Historically, the white-dominated medical profession, with its natural interest in bodies and minds, contributed extensively to the construction of racialized hierarchies by producing ideas about people's social status formulated around skin color. The therapeutic professions actively contributed to the processes of defining who was "inferior" and who was "superior." In relation to African Americans, doctors had discussed "negro inferiority" since at least the eighteenth century, and postwar psychologists emulated these hierarchical presumptions while assessing Black inner life.[10] This formed part of a far longer tradition in which supposedly objective science (created historically largely by white middle-class men) implicitly advanced ideas of women and other groups as inferior.[11] Unconscious data bias across medicine and science has been (rightly) criticized for supporting a world built for, and by, men.[12]

Activists have always trodden carefully when broaching the issue of inferiority. To claim that inequality and discrimination cause emotional damage (with low self-esteem being a key symptom) risks furnishing further evidence of inferiority, inability to cope psychologically with modern life's demands, and unsuitability for social equality. Highlighting weakness could arm conservatives and racists with further justification for exclusion tactics. While claiming emotional harm, activists have therefore found it important to simultaneously express confidence, pride, and self-esteem, and, at the same time, acknowledge a traumatized past.[13] By necessity, Black pride and Black misery had to coexist.

The offensive idea of "negro inferiority" is centuries old. From the late nineteenth century, eugenicists propagated ideas that nonwhites were biologically inferior (and also, by extension, morally and culturally inferior). Social problems in that group were rendered innate to the "Negro," an assumption that overlooked the role of socioeconomic environments rife with discrimination. Western biology, which depicted all races as having inborn characteristics, was developed almost exclusively by white, middle- to upper-class men who portrayed anyone outside that narrow group as inferior.[14] Adler's "inferiority complex" acquired considerable significance in the politics of the marginalized.

Not that we should entirely dismiss all aspects of "negro inferiority" as artificial caricatures. Many African Americans probably did endure a poor self-image, given the racist contexts in which they lived, although these would not be due to any innate deficiency. In the early twentieth century, prolific Black intellectual W.E.B. Du Bois influentially wrote that people of color were subjected to a double consciousness, or two conflicting identities, "a painful self-consciousness, an almost morbid sense of personality, and a moral hesitancy which is fatal to self-consciousness." This double consciousness severely undermined self-esteem. Blacks were at once American and negro, "two warring ideals in one dark body," although they no longer wanted to feel rejected.[15] Early twentieth-century leaders such as Mary Church Terrell, co-founder and first president of the National Association of Colored Women, also advocated a positive Black self-image, seeing this as an important basis for a dignified future.[16]

Particularly after World War II, social scientists urgently rethought race theory. The Holocaust's architects had applied extreme biological theory to instigate acts of extermination. Researchers who once waxed lyrical about the wonders of eugenics swiftly retreated, at least publicly. Even so, race, as an idea, remained firmly intact, despite its biological reality, which was now considered questionable. Race remained incorporated into everyday language.[17] In the 1940s, calls were made to "explode the concept of negro inferiority."[18] Some liberal social scientists insisted that the idea had never been empirically proven and dismissed it as blatantly racist.[19] Nonetheless, rather than being entirely discarded, the premise of "negro inferiority" was redefined in psychological terms. Therapeutic culture's rising influence underpinned this rethinking of both race and racism.

Following the war, research into Black psychology boomed. Much of this investigated racism's psychological impacts, a compassionate departure from earlier preoccupations with delineating racial superiority and inferiority. It sought to understand the personality types being forged in discriminatory environments. Psychology's influence was also felt in Black popular culture. Widely read novels such as Richard Wright's *Native Son* (1940) and his memoir *Black Boy* (1945) further helped to move discussion toward emotions and psychological wellbeing. Openly inspired by psychological theory, Wright penned systematic studies of the mental habits, dispositions, mechanisms, and practices rife in racist cultures, critiquing white supremacy by interrogating the emotional realm. Prominent writer James Baldwin, too, enticed Black Americans to see more clearly how their interior lives were connected to the general condition of a country that valued whites far more than Blacks. Mind and country were inextricably related.[20]

Until the 1960s, psychologists took it for granted that Black people had lower self-esteem than whites. Blacks were exposed constantly to racism and socioeconomic inequality, so this seemed believable. Psychologists, supportive or otherwise, presumed that prejudice destroyed self-image.[21] Even well-meaning, sympathetic psychological texts of the time perpetuated typecasts of the "damaged negro," a subjugated individual with an inherently flawed psychological life, one in constant emotional turmoil due to inexorable feelings of inferiority. The genre of (usually) white-authored studies of "negro personality" published in the 1950s and early 1960s revealed more about the white liberal personalities who wrote them than about African Americans themselves. Rather than redeeming Black personalities, liberal white psychologists, even when genuinely trying to help, still caricatured Black Americans as a homogenous group perpetually troubled by emotional distress and personalities developed entirely in relation to white racism.[22] This did not go unnoticed. Ralph Ellison, famed author of *The Invisible Man* (1952), observed that researchers, when examining "the negro," dropped their critical armament to "revert with an air of superiority to quite primitive modes of analysis."[23] They racialized Black culture by offering dismal images of Black people hindered by a nagging inability to feel good about themselves and possessing limited psychological variation.[24] This was not a self-image to be proud of.

A well-known example of this genre is Abram Kardiner and Lionel Ovesey's influential 1951 book, *The Mark of Oppression: A Psychosocial Study of the American Negro*, a psychodynamic study based on twenty-five individuals. Interested in the psychoanalytic study of culture, the authors portrayed Black psychology as oriented almost entirely around discrimination and self-hatred. As Blacks constantly filtered hostile images of themselves from whites, their self-esteem dropped to worrying levels. Kardiner and Ovesey portrayed social discrimination as "an ever-present and unrelieved irritant," "painful in its intensity," as something that hindered the Black person's ability to "maintain internal balance and to protect himself from being overwhelmed by it." Black people were always maneuvering, emotionally, to keep functioning, constantly adapting to preserve social effectiveness in the face of adversity.[25]

The self-hatred wrought by discrimination, marginalization, and alienation took center stage in this psychological redefining of race. The logic ran that whites treated African Americans as inferior on the basis of skin color. Blacks then internalized these negative perceptions, lost their self-value, and began looking upon themselves with loathing and disgust. Sociologist Everett V. Stonequist was an early influence. In the 1930s, he developed the concept of "the marginal man," an individual who experienced group conflict as a personal problem. According to Stonequist, marginal men absorbed the culture of the dominant group while young, still unaware that they were not, and could never be, part of that group. The moment of realization of being a permanent outsider, and of feeling marginalized, marked a significant psychological turning point. The sudden, unexpected shock disorganized the emotions; negative personality traits began to form.[26] Kardiner and Ovesey also believed that self-hatred caused Blacks to idealize and revere whiteness. An unattainable desire to become white manifested itself in hostility toward whites, hatred of other Blacks, apathy, crime, and hedonism. Black people's subconscious desire to be white fueled an emotional world of frustration and self-hate. The greater the idealization of whiteness, the greater the intrapsychic discomfort, "for he [the Black person] could never become white."[27]

Not that this stopped some Black people from trying. Social scientists had long been interested in African Americans who try to "pass" as white, intentionally or unintentionally, permanently or temporarily, fully or partially. Earlier studies focused on socioeconomic motivations. Pretending to be white might help secure work, providing one's skin was pale enough.[28] In *The Marginal Man*, Stonequist described "passing" as an effort by the marginalized to assimilate into the dominant social group. But unlike, say, white Jewish or gay communities, the basic, and obvious, fact of different skin color made "passing" trickier for most African Americans.[29] In contrast, postwar social scientists developed explanations rooted in psychology and emotions. They reinterpreted "passing" as a psychological defense mechanism, not a socioeconomic necessity, a tell-tale sign of self-hatred and self-rejection.[30]

Kardiner and Ovesey's *The Mark of Oppression* was remarkably influential, with its persuasive application of psychodynamic approaches to understanding racism, then a pressing social matter. Ultimately, though, it affirmed woeful portrayals of Black life that pathologized "negroes." It retained typecasts of matriarchal African American families and Black men plagued by anxieties about their masculinity.[31] It contained derogatory images of suffocating Black mothers as a key cause of the Black man's oppression.[32] Not that the book deliberately intended to be racist, but it nonetheless wound up offering depressing images of African American personalities that intimated white superiority. While exposing racism's harmful psychological consequences, the authors unwittingly caricatured emasculated, overdependent, depressed, self-hating Black people. In that sense, *The Mark of Oppression* was curiously ethnocentric.

Other notable white liberal research includes that of Thomas Pettigrew, a social psychologist with genuine civil rights sympathies. In 1964, he wrote that, "for years, negro Americans have had little else by which to judge themselves than the second-class status assigned to them in America. And along with this inferior treatment, their ears have been filled with the din of white racists egotistically insisting that Caucasians are innately superior to negroes." Like other psychologists, Pettigrew believed that most Black people had absorbed ideas about their own inferiority.[33] Robert Coles's monumental 1967 study, *Children of Crisis*, examined African American strategies used to offset the negative effects of discrimination, such as positive coping strategies.[34] However, these social scientists also struggled to move away from the trappings set in *The Mark of Oppression*.[35] Painting a picture far too rosy would have limited opportunities to express the severity of Black emotional problems. Many critics dismiss these social scientists as researchers who were unable to fully escape their racist upbringing, even though, in reality, they actively supported civil rights and intended their damaged imagery to be read as a moral critique of American society. Despite later acquiring a reputation for being racist, Ovesey was inspired by his parents' belief in social justice; he attended civil rights marches, and stated that Kardiner too generally agreed with this worldview.[36]

Black and white dolls

To some extent, the research outlined above can be interpreted as whites "gazing" on Blacks. Philosophers speculate that the "white gaze" objectifies Black bodies (and minds) as entities to be feared, disciplined, marginalized, and segregated. White bodies are cast as normative and Black bodies as pathological through a "distorted way of seeing" that ultimately stemmed from myths about race and processes of white supremacism.[37] Such ideas certainly apply to the studies just discussed, which cast the Black psyche as weak, pathogenic, and deficient. However, relying on supportive white social scientists was useful only to a certain extent. One of these, Robert Coles, openly acknowledged that his white, middle-class background precluded an intuitive access to knowledge of Black communities.[38] African Americans needed a voice of their own.

In the 1950s, Black self-esteem became nationally discussed. Kenneth B. Clark's *Effect of Prejudice and Discrimination on Personality Development* was one of seven social science works examined for a Supreme Court investigation into school segregation.[39] Clark was a prominent African American psychologist and civil rights campaigner. In his report, later republished for a general audience, he insisted that white children weren't born with racial feelings; they learnt these in a culture that was supportive of racism. Clark thought he had irrefutable evidence that, because they felt rejected, most Black children would prefer to be white.[40] Famously, he presented this at the landmark 1954 Supreme Court case *Brown v. Board of Education*, which dealt a lethal blow to the educational segregation policies established in 1896 by *Plessy v. Ferguson*. Clark's evidence, compiled with social psychologist (and his wife) Mamie Clark, was based on an experiment that involved asking children to choose between brown and white dolls. The majority of Black children saw their resemblance in the brown dolls, but nonetheless chose to play with a white doll, seeing its skin color as more pleasing. For Clark, here was firm proof that Black children internalized white racism. The methodology was later discredited, but the research still holds a hallowed place in social psychology, and its significance at the time should not be underestimated.[41]

Black self-esteem was suddenly a very public matter. Clark's arguments that racial segregation damaged Black children's self-esteem, usually permanently, proved convincing.[42] The brief read: "To separate [African American children] from others of similar age and qualifications solely because of their race generates a feeling of inferiority as to their status in the community that may affect their hearts and minds in a way unlikely ever to be undone."[43] The proceedings are also notable for foregrounding psychological evidence as a focal point of debate about race, highlighting the increasing centrality of therapeutic ideals to postwar racial liberalism.[44]

Racial representation was complex. Negative representations left some Blacks feeling downtrodden, but others found dynamic ways of taking control. Clark seized on opportunities to present African American life as pathogenic for its potential to challenge racism and elicit some white sympathy. As psychology at the time generally focused on negatives, Clark's research fitted well into the discipline's burgeoning positive turn.[45] Clark was willing to agree with white liberal assessments of Black self-esteem since they helped expose the inner trauma wrought by racism. Ultimately, *Brown v. Board of Education* didn't transform American education overnight. Indeed, lawyer, professor, and civil rights activist Derrick Bell describes its "unfulfilled hopes for racial reform."[46] Nonetheless, it was hugely symbolic.

Throughout the 1950s and 1960s, Clark continued to argue that Black children believed that whites viewed them unfavorably, lowering their self-esteem. He thought that early experiences were difficult to overcome, "for the child who never learns to read cannot become a success at a job or in a society where education and culture are necessary." Black teenage pupils suffered from low motivation and gradually withdrew. Poor-quality teaching in ghetto schools reinforced this sense of inferiority, leading later in life to cycles of menial jobs, "broken homes," and prison spells. As adults, they were unmotivated, uninterested in civic affairs, lethargic when it came to voting, and suffering from an "irresponsibility rooted in hopelessness."[47] A characteristic summation can be found in Clark's 1965 book *Dark Ghetto*:

> Human beings who are forced to live under ghetto conditions and whose daily experience tells them that almost nowhere in society are they respected and granted the ordinary dignity and courtesy accorded to others will, as a matter of course, begin to doubt their own worth. Since every human being depends upon his cumulative experience with others for clues as to how he should view and value himself, children who are consistently rejected understandably begin to question and doubt whether they, their family, and their group really deserve no more respect from the larger society than they receive. These doubts become the seeds of a pernicious self- and group-hatred, the Negro's complex and debilitating prejudice against himself. The preoccupation of many Negros with hair straighteners, skin bleachers, and the like illustrates this tragic aspect of American racial prejudice. Negroes have come to believe in their own inferiority.[48]

Clark's assessment of Black self-esteem vividly depicted a process of what sociologists call "psychological colonization" (although it was not overtly framed in this way) in which identity, not freedom, economic success, or political rights, was lost and denied. His approach reframed civil rights issues in direct relation to the self. Nonetheless, the caricaturized "damaged negro" image persisted, a person too plagued by self-hatred and inferior feelings for white people to let them integrate. While this approach undoubtedly had practical purposes, Clark's conclusions still correlated problematically with the "mark of oppression" model.[49]

Then again, rather than be entirely critical, we must bear in mind that damage imagery did effectively call attention to very real emotional problems rife across Black communities. To explore an autobiographical source, in 1993 Julia A. Boyd recollected the first time her mother explained why white and Black girls had different hair. Boyd recalled this as a significant, life-changing childhood moment, one in which she became aware of her difference from others. While young, she grew up with very few positive images of Black women. Those few available tended to be printed in magazines aimed only at an African American readership, such as *Ebony* and *Jet*. Even these scarcely reflected the diversity of Black women, thought Boyd.[50] Despite adoring the *Gone with the Wind* film, it nonetheless made her aware that Black women as heroines were absent from her favorite films and books, constraining her potential to see herself as a confident, powerful Black woman. Boyd was convinced that these childhood incidents impacted negatively on her sense of self-esteem, as social messages became personal beliefs about herself. In her own words: "These distorted messages are repeatedly reinforced through the media, in our personal contacts, sometimes even in our families. When we internalize these messages, they cloud and poison our self-esteem. It's like eating bad food or breathing toxic air. We get sick."[51]

Black rage

From the late 1960s, Black activists banded together in protest groups and social movements, inspiring positive feelings of pride and self-confidence. They rejected pessimistic psychological depictions, even accusing white liberal researchers of racism. As Ibram X. Kendi observes, Clark's use of the phrase "dark ghetto" was intended to broadcast the sense of ruthless segregation and poverty facing urban Blacks, but the phrase quickly took on a racist life on its own.[52] More positive affirmations of Black life were needed.

Much of the impetus for a Black-led interpretation of African American psychology came from French West Indian psychiatrist Frantz Fanon, who, in 1952, had published a hugely influential book, *Black Skin, White Masks*, outlining psychoanalytical interpretations of the "Black problem." Drawing influence from Freud and Adler, and rejecting ideas on the fixed nature of racialized personalities, Fanon argued that inferior feelings among the colonized were the correlative of white feelings of superiority, for "it is the racist who creates his inferior." Fanon wanted Black people to take action, individually and collectively, to became aware of their unconscious feelings, and to abandon efforts to become white. "The Black man should no longer be confronted by the dilemma 'turn white or disappear'," he wrote, "but he should be able to take cognizance of a possibility of existence." Fanon added that Black people should stop experiencing themselves through the lens of others, especially those who disliked them "for not only must the Black man be Black, he must be Black in relation to the white man."[53] Fanon wanted African Americans to reject claims about their inferiority and assert pride and self-confidence.

There were signs that Black self-esteem levels might be changing. In *The Negro's Morale* (1949), (white) sociologist Arnold M. Rose argued that Black group identification had strengthened during the 1940s, and that "negroes also have a great growth of race pride. In fact, some of them now have the trait in pathological form. For some, the glorification of the negro group is considered to be the main purpose of negro activity. Radicalism and nationalism have become associated with race pride." If true, African Americans were recapturing a sense of confidence, pride, and esteem.[54] In 1953, sociologists George Eaton Simpson and J. Milton Yinger (both also white) wrote: "A person with good ears will hear, we believe, genuine expressions of pride and self-confidence and a reciprocal dismay at some of our society's foibles."[55]

Of course, in reality, many African Americans may never have been truly reconciled to their subordinate status or uncritically accepted the naturalness and permanency of white supremacy. Historically, prominent Black leaders had rejected ideas of Black inferiority, including Booker T. Washington and W.E.B. Du Bois.[56] Similarly, it would be wrong to presume that such changes occurred overnight from the 1950s. From at least the 1940s, even white researchers recognized that Black community groups, newspapers, humor, and group solidarity encouraged positive self-esteem among African Americans.[57] Nonetheless, from the 1950s, Black self-confidence became far more visible and politically charged.

Later, a new narrative emerged. This time, it pinpointed the precise moment when Blacks recovered their self-esteem: when Rosa Parks famously refused to sit alongside other African Americans at the back of a Montgomery bus in 1955. When interviewed, Parks herself portrayed this as a symbolic turning point, stating that "it was a matter of dignity. I could not have faced myself and my people if I had moved." "What I learned best," Parks remembered, "was that I was a person with dignity and self-respect, and I should not set my sights lower than anybody else just because I am Black."[58] The Montgomery bus incident is still upheld as a historic turning point, when African Americans realized that many stereotypes about themselves weren't actually true. As one author wrote in 1965: "In Montgomery, we walk in a new way. We hold our heads in a new way."[59] Jeanne Theoharis, on the other hand, has succinctly written about the way in which Rosa Parks has been iconized. She describes a "myth of Parks" that has largely overlooked her six decades of activism and militant radicalism, concentrating instead on an image of unassailable respectability.[60] In reality, Parks was just one of many southern Black women prominently involved in mass demonstrations and civil disobedience.[61] Indeed, she wasn't even the first Black person to refuse to give up her seat, although she was perhaps one of the most respectable looking of the women who did so.[62]

By the late 1950s, it was no longer necessary to have "good ears" to hear the rising confidence of some African Americans. Another oft-cited symbolic moment occurred in 1959, when Malcolm X became a household name in America after the television broadcast of *The Hate That Hate Produced*. In this, Malcolm X charged the white man with being the biggest liar, drunkard, murderer, peace-breaker, robber, deceiver, and trouble-maker on Earth. During a mock morality play called "The Trial," the white man was metaphorically sentenced to death for his crimes against African Americans. During the course of the program, shocked audiences became acquainted, many for the first time, with the Nation of Islam, "the most powerful of the Black supremacist groups."

Many white viewers were stunned to learn that Black people harbored such anger toward them. Within weeks of the broadcast, the Nation of Islam's membership doubled to 60,000. The underlying rage of African Americans had manifested in a threatening political assertiveness.

This fed into ideas prevalent in damage imagery texts that warned of the underlying rage contained within the pathologized, interiorized "negro." Psychologists cautioned that just below the surface of the "negro's mask of compliance" simmered anger and rage.[63] Kardiner and Ovesey spoke of the Black psyche's subconscious domination by rage stemming from low self-esteem and self-hatred, and constant efforts to suppress and curtail this fury. While attempting to control this rage, resentment built. The result? "A personality devoid of confidence in human relations" and "an external vigilance and distrust of others." According to *The Mark of Oppression*, the root of the problem was the lingering impression that "I am not a lovable creature," that "no-one could love a negro."[64] In *Native Son*, Wright further developed this Black rage thesis. He portrayed serious crimes such as rape and murder, when committed by Black men, as a consequence not of inherent sociobiological defect but of the psychological conditions wrought by white racism. Wright's thesis had its origin in William James's writings, who himself wrote of how bottled rage and despair could overwhelm those treated harshly. This compelling idea was adaptable to the postwar Black situation.[65]

Malcolm X's emphasis on racial pride was not entirely new. It had been central to Jamaican activist Marcus Garvey's popularity in the 1920s. However, X's interpretation of pride was notably militant. He insisted that Black Americans needed to purge themselves of a false consciousness that, through slavery and racism, had distorted their personalities.[66] To gain a sense of Malcolm X's overt hatred of white America before his 1964 arrival at Mecca, one could refer to the following passage from his autobiography: "I was among the millions of Negroes who were insane enough to feel that it was some kind of status symbol to be light-complexioned, that one was actually fortunate to be born thus. But, still later, I learned to hate every drop of that white rapist's blood in me." He continued:

> I actually believe that as anti-white as my father was, he was subconsciously so afflicted with the white man's brainwashing of Negroes that he inclined to

favor the light ones, and I was his lightest child. Most negro parents in those days would almost instinctively treat any lighter children better than they did the darker ones. It came directly from the slavery tradition that the "mulatto," because he was visibly nearer to white, was therefore "better."[67]

Integrationists such as Martin Luther King were similarly cognizant of widespread Black self-esteem issues. In his 1967 book, *Where Do We Go from Here? Chaos or Community*, King argued that "psychological freedom, a firm sense of self-esteem, is the most powerful weapon against the long night of physical slavery." This type of freedom could only emanate from the mindset of Blacks, rather than from a civil rights bill. Asserting positive selfhood would allow African Americans to become truly free. King added:

> The Negro must boldly throw off the manacles of self-abnegation and say to himself and the world: "I am somebody. I am a person. I am a man with dignity and honor. I have a rich and noble history, however painful and exploited that history has been. I am Black *and* comely" ... This is positive and necessary power for Black people.[68]

However, Malcolm X, and later the Black power movement, added a new sense of urgency to the ongoing discussion. They weaponized their negative inner feelings. While aspiring to positive emotions such as happiness and self-esteem, they warned at the same time of the underlying rage that generations of emotional damage had created. Indeed, perhaps the expression of Black rage was itself a useful form of therapy.

Black power

In the late 1960s, the Black Panthers questioned, undermined, and subverted psychological discourse on "negro inferiority," replacing this with their own interpretations. Many Black people had grown impatient of the integrationists' peaceful, conciliatory approaches. Some saw assimilationist ideas as problematic, even essentially racist, as they upheld whiteness as a superior standard into which Black people should integrate.[69] As chair of the Student Nonviolent Coordinating

Committee, Stokely Carmichael, alongside King, participated in the Mississippi March Against Fear in the summer of 1966, launched by James Meredith. It was there that he first used the phrase "Black Power." Subsequently, Carmichael traveled extensively across the nation building up his movement. The rapid growth of the Black Panther Party in 1967 suggested that Black youth saw civil rights strategies as failing to have the desired effects. No matter how many peaceful demonstrations were staged, issues of police brutality, institutional racism, poor housing, inferior schools all continued to plague Black working-class communities.[70] Among its many aims, the themes of pride and self-esteem stand out for the purposes of this book.

In 1967, Carmichael and Charles Hamilton called for a new consciousness among Black people based on pride, not shame. They critiqued integration and provided new ways of using psychological language to articulate Black feelings. In their book *Black Power*, they quoted E. Franklin Frazier's *Black Bourgeoisie*, as well as Stonequist's *Marginal Man*, to confirm that Black self-esteem had been historically depleted.[71] However, there was now a twist: African Americans developed their own sociopsychological philosophy and redefined racial representations on their own terms. Unlike the sympathetic white liberal psychologists, Carmichael and Hamilton urged African Americans to reclaim their history and identity from "cultural terrorism."

A related line of argument insisted that white racist society was "sick" and "pathological," a victim of its own racism.[72] Carmichael and Hamilton argued that white guilt about their own racism was problematic.[73] Some white liberal psychologists agreed. In 1971, Baughman wrote that Blacks overestimated the extent of self-esteem possessed by white people; indeed, the white person's constant need to engage in racist behavior hinted at psychological fragility and insecurity. "If in fact he felt secure and adequate," wrote Baughman, "such behavior would not be necessary."[74] African American psychiatrist Alvin Pouissant proposed a similar argument: that the idea of Black self-hatred allowed whites to feel that while African Americans were psychologically deranged, they themselves were models of mental health. "In fact," suggested Pouissant, "it must be whites who are insecure and filled with self-hatred."[75]

Carmichael and Hamilton insisted that African American racial and cultural personality needed to be preserved, not integrated or made whiter. Their interpretation of Black personality was rooted in pride, defined as self-acceptance (but not chauvinism) in being Black and appreciating the historical achievements and contributions of African Americans. As they wrote: "No person can be healthy, complete, and mature if he must deny a part of himself." Carmichael and Hamilton's rejuvenation of African American life placed issues such as self-esteem and self-acceptance at the very core of their philosophy, encouraging group improvement of the self based on pride.[76] But they failed to precisely outline ways to build positive self-esteem; this was not a self-help book per se. Nevertheless, its basic premise was powerful: building a positive self-image was pivotal to Black liberation. The authors wanted, in their words, to "establish a viable psychological, political, and social base upon which the Black community can function to meet its needs."[77] Carmichael's popularization of the Black power slogan marked a new stage in the transformation of Afro-American consciousness, one unafraid to articulate Black rage.

The Black Power movement saw potential in African American cultural regeneration, a potent symbol of Black separatism. Cultural nationalists argued that white oppression had turned Blacks into people enamored of everything white and repulsed by their own skin color. Offering an alternative, Black Power rejected the very idea of being "negro," and instead proposed celebrating Blackness. They urged Blacks to reject white culture and become "African Americans." The use of African names, clothes, rituals, and sensibilities helped reconnect Black culture to that of a (somewhat idealized) African homeland.[78]

Black Power found its own expert in Poussaint, who was not the usual ivory-tower psychiatrist. Before engaging with Black Power, he had already been closely involved in the civil rights movement. In 1965, he moved to Jackson, Mississippi, and worked for a year as Southern Field Director of the Medical Committee for Human Rights. In this role, he fought to end racial discrimination in southern health facilities. After the passing of the Civil Rights Act of 1964, he helped desegregate the hospitals. At the 1965 Selma to Montgomery marches, Martin Luther King reportedly took Pouissant aside to thank him for his valuable observations about Black psychological life.[79]

Writing in 1968, Poussaint looked back to the emergence of the Black Muslim Movement, not to Rosa Parks, as a turning point for Black identity. He conceded that many African Americans suffered from a tarnished self-image, and that whites had socialized African Americans to "internalize and believe all of the many vile things he said about him," encouraging behavior and attitudes that reaffirmed harmful stereotypes. In his words: "Our mass media disseminated these images with vigor on radio, in movies ... and like unrelenting electric shocks conditioned the mind of the Negro to say 'Yes, I am inferior'." He continued:

> Black men were taught to despise their kinky hair, broad nose and thick lips. Our "black" magazines pushed the straightening of hair and bleaching cream as major weapons in the Negro's fight for social acceptability and psychological comfort.
>
> The most tragic, yet predictable, part of all of this is that the Negro has come to form his own self-image and self-concept on the basis of what white racists have prescribed.

Poussaint saw Black Power as radical advocates of a new self-image. He was impressed by their efforts to rehabilitate self-esteem and alleviate "Negro self-hatred without holding up or espousing integration or 'full acceptance' of the black man into American white society."[80]

As a psychiatrist, Poussaint believed that self-help was crucially important to Black Power: "Black people must undo the centuries of brainwashing by the white man, and substitute in its stead a positive self-image and positive concepts of oneself – and that self happens to be the black, dispossessed, disenchanted, and particularly poverty-stricken Negro." As he also wrote: "The major source of the Black man's twisted self-image is the racism in our society. Being powerless and poverty stricken has contributed more to his lack of self-esteem than has his skin color."[81]

Activists robustly rejected "passing." In 1964, novelist and civil rights activist John O'Killens published an article in the *New York Times* in which he lambasted the practice. Instead, he argued, Black blood and intelligence should be injected into mainstream American life to revitalize it. Critical of the potential for integration to encourage African Americans to adopt white lifestyles, he wrote: "We respect ourselves too much for that." For O'Killens, freedom to be Black meant living in an equal society while being able to retain and express, without fear, a strong Black identity. "This is the only way that integration can mean dignity for both of us," he claimed, when differentiating between "racial sameness" and "racial equality." He concluded: "Wyes, we are different from you and we are not invisible men, Ralph Ellison notwithstanding ... We are both Americans and Negroes."[82]

These activists were acutely aware of the damage inflicted by racist social environments on emotional wellbeing, and how racism and capitalism converged to damage minds. They were therefore more willing to consider the broader structural determinants of emotional ill-health. They dreamt of social change, as did humanistic psychologists, but their idea of happiness was a world in which they were accepted. Black people had clearly defined obstacles that needed to be removed before they could find their own form of happiness in America.

The Black Panthers saw psychological health as a basic human right and furiously debated its relation to racism.[83] Black consciousness programs were intended to build self-confidence and encourage African Americans to think and do for themselves. Problem-solving and community leadership were intended to undo the Black people's learned self-hatred and self-destructive behavior. At the grassroots level, projects in places such as Louisiana focused on self-help schemes that encouraged African Americans to take care of their neighborhoods, obtain better jobs, and reduce school drop-out levels. In 1965, the *Wall Street Journal* even suggested that "Self-help" might eventually replace "Freedom Now" as a major emphasis in civil rights activities.[84] At the micro level, expressing self-esteem and pride was a symbolic act of subversion in which they could all potentially partake. Robin D.G. Kelley points out that African Americans developed daily acts of clandestine resistance.[85] It seems to me that simple self-help strategies, such as expressing, and preferably also feeling, self-esteem and pride, were among these subtle, but incendiary, acts embraced by Black communities from grassroot to leadership level.

A proud Black culture

Across the nation, African Americans re-created their physical appearance, visibly inscribing feelings of pride and self-confidence onto their body.

Hair was crucially important. In 1964, O'Killens, mirroring Kenneth Clark's aforementioned criticisms, wrote:

> I work for the day when my people will be free of the racist pressures to be *white like you*; a day when "good hair" and "high yaller" and bleaching cream and hair straighteners will be obsolete. What a tiresome place America would be if freedom meant we all had to think alike and be the same color and wear the same grey flannel suit![86]

In light of this, a new Black aesthetic was proposed that expressed pride in "negroid" features. Putting on display natural Black hair, the Afro, symbolically embraced African heritage. Young civil rights campaigners wore their hair natural and unprocessed to express Black solidarity and authenticity, repudiating an oppressive white aesthetic.[87]

Hair was surprisingly politicized and fed into psychological analyses of Black self-esteem. In his autobiography, Malcolm X lamented the days when he had adorned a conk:

> The ironic thing is that I have never heard any woman, white or Black, express any admiration for a conk. Of course, any white woman with a Black man isn't thinking about his hair. But I don't see how on earth a Black woman with any race pride could walk down the street with any Black man wearing a conk, the emblem of his shame that he is Black.
>
> To my own shame, when I say all of this I'm talking first of all about myself, because you can't show me any Negro who ever conked more faithfully than I did. I'm speaking about personal experience when I say of any Black man who conks today, or any white-wigged Black woman, that if they gave the brains in their heads just half as much attention as they do their hair, they would be a thousand times better off.[88]

Similarly, prominent Black author Ta-Nehisi Coates later recalled of this period:

> My mother, at twenty, let her relaxer grow out, and cultivated her own natural, nappy hair. The community of my youth was populated by women of similar ilk. They wore their hair in manifold ways – dreadlocks and Nubian twists, Afros as wide as planets or low and tapered from the temple. They braided it,

invested it with beads and yarn, pulled the whole of it back into a crown, or wrapped it in yards of African fabric. But in a rejection aimed at something greater than follicles and roots, all of them repudiated straighteners.[89]

Black music, and its many stars, contributed greatly to the public visibility of the new self-image, one filled with just as much self-confidence as the idealized white person's personality, but one that was at the same time uniquely Black. During the civil rights period, Black musicians thought it important to appear respectable. Suited Black soul singers sang harmonic ballads; Diana Ross and the Supremes wore elegant cocktail dresses. They self-consciously fashioned their bodies to project an aura of Black respectability, implying their worthiness of equal citizenship. However, this bodily refashioning gave way later to bushy afros, big jewelry, and raised fists, heralding the decline of respectability as a political strategy.[90]

Blues music was originally conceived by ex-slaves and freedman. For 1960s activists, it seemed too reminiscent of a past era of hardship and sadness, offering little hope of future improvement.[91] In the early 1960s, Charles Keil chose blues as the subject of his anthropological study of Black culture.[92] But the blues didn't suit the climate of rising Black self-confidence; jazz seemed more suitable as the soundtrack for Black liberation,[93] while funk, soul, and rhythm and blues provided further soundtracks. James Brown abandoned his conked, chemically straightened hair, sported by most male rhythm and blues singers until the mid-1960s. His Afro and pro-Black songs articulated the spirit of the time and also Black Power's influence on popular culture.[94] Brown was fortunate to be successful enough to do what he pleased, since he owned how own production company.[95] In his popular 1968 song *Black and Proud*, he preached the message that "Black folk need to respect themselves." A funk song released in 1968, it was noticeably more shouty and aggressive than earlier civil rights movement songs (such as the sedate *We Shall Overcome*). It was self-consciously less respectable but at the same time upbeat and positive. The song's call-and-response chorus was performed by a group of young children who responded chirpily to Brown's lyric "Say it Loud!" with "I'm Black and I'm proud."[96]

Changes in Black culture also influenced literature. In his earlier writings, James Baldwin had written regularly about saving the souls of white people by warning them of the consequences of failing to change. In the radicalized late 1960s, he focused more on the wellbeing of Black people than on white America, mirroring the changing tactics of Black militancy.[97] Film, too, now featured supremely self-confident Black characters in leading roles. The controversial blaxploitation genre emerged in response to stereotypical misrepresentation of Black culture in mainstream films. Black cinema-goers were tiring of watching Black characters trying to fit into the white world. Sidney Poitier, for example, fell out of favor.[98] Despite being an anti-Semitic, homophobic killer, the fictional private detective John Shaft presented African Americans with a new type of hero previously denied to them. The boxer Muhammed Ali became a world-leading sportsman, despite his skin color. Black Panther member Eldridge Cleaver wrote that earlier Black boxing champions had been puppets manipulated by whites. In contrast, Ali refused to be attached to the white man's strings. Cleaver wrote: "For every white man, feeling himself superior to every Black man, it was a serious blow to his self-image; because Muhammed Ali, by the very fact that he leads an autonomous private life, cannot fulfil the psychological needs of whites." Ali provided a potent, and positive, symbol of Black masculinity, causing whites, Cleaver believed, to suffer an identity crisis.[99]

However, it would be misleading to write an entirely optimistic account. Black Power had complexities and ambiguities that were often overlooked in hagiographic retellings.[100] For the most part, this chapter has spoken of Black people, male and female, in quite general terms. In reality, male activists urged women to eschew positions of leadership and authority and stay at home. As Jacqueline Jones succinctly puts it: "Too often male Black Power advocates, from activists to scholars, wanted their female kin to get out of the white man's kitchen and back into their own."[101] With the exception of Parks, women are too often excluded from the dominant Black narrative, despite the fact that many women activists embraced racial separatism, Black pride, and self-determination. They were also notably active in grassroots initiatives. Black women, some say, formed the backbone of the movement.[102]

The gendered complexities of Black Power were also manifest in Cleaver's writing, published while acting as the Black Panther Party's minister of information. His 1968 book *Soul on Ice* provided a manifesto of Black Power masculinity intended to redeem the tragic colonized male, whose soul was "on ice," whose being was the "Black eunuch." Like Fanon, Cleaver, writing in prison, wrote about decolonizing Black souls. After a number of religious experiences in prison, he became a Muslim convert, preacher, and follower of Malcolm X. Through this process, he believed that he regained his previously alienated and splintered self-image accrued when growing up as a child in a Californian ghetto. From this time onward, he began the process of self-analysis, self-education, and self-expression. Cleaver recalled his youth, before finding religion, describing himself as a rapist, as practicing on Black girls in the ghetto, a place that no one from the outside cared much about what was happening, before defiling white women against their will as revenge on "the white man's law." Later in life, Cleaver lost his self-respect after coming to regret his violent youthful behavior. In his words: "My pride as a man dissolved and my whole fragile moral structure seemed to collapse, completely shattered. That is why I started to write. To save myself."[103]

To say the least, Cleaver was dismissive about some of his prominent African American contemporaries. He portrayed Baldwin as a self-hating Black man who, upon discovering the inaccessibility of his African heritage, had appropriated the white man's heritage and made it his own even though, in doing so, Baldwin revealed himself to hate and fear white people. Cleaver dismissed the integrationists as driven by an ethnic self-hatred, which he described as a "racial death wish."[104] Cleaver's misogynism wavered little after his religious conversion. He described Black women as race traitors and blamed them for emasculating Black men. As bell hooks later declared, labeling women this way was potentially devastating to the self-esteem of Black women who might have benefited from the movement's emphasis on creating positive self-images. To be told that their efforts to end racism were emasculating and damaging to the cause must have felt extremely confusing.[105]

Tensions also arose between the peaceful shaping of a positive self-image brimming with pride, respect, and esteem, and the violent riots across disaffected Black communities. The summer of 1967, in particular, was one of race riots and police brutality. In a *New York Times* interview, Clark warned that Black people were "affirming their power to destroy. In the act of rebelling, they appear to be asserting the ability to rebel." Beneath this destructive and irrational behavior was the "pathetic logic of asserting self-esteem and searching for a positive identity by exposing oneself to danger and even inviting death." For Clark, "this is the quest for self-esteem of the truly desperate human being. This is the way those who have absolutely nothing to lose seek a pathetic affirmation of self even it is obtained moments before death." He insisted that, because of the urgent injustices being inflicted upon them, Black youths didn't have the luxury of being able to "escape into the oblivion of hippie-land with the same seeming equanimity of their white counterparts," adding that "one must understand that no group of human beings can move from being the victims of extremes of injustice and inhumanity to the goals of self-acceptance and positive personal and racial identity without a transition period being marked by turmoil."[106]

For Clark, the turn to the therapeutic was not a calm, linear path to happiness for disaffected African Americans. For a start, unlike white counterculture activists, Black rage had to be played out first. And it is here where we can see perhaps most clearly what differentiated the harnessing of self-esteem, psychology, and self-help by Black activists. Of course, it could also have been the case that many angry Black activists saw therapeutic approaches as too passive and lacking in urgency, perhaps even too white. For little compared to a riot in its ability to grab attention.

Did Black self-esteem improve?

So did Black self-esteem increase during the 1960s? This is difficult to ascertain, retrospectively. From the late 1960s, contemporary psychologists scrutinized Kardiner and Ovesey's *The Mark of Oppression* to expose its reliance on just twenty-five case studies to support sweeping views on Black life.[107] Keil thought that psychologists, specifically Kardiner and Ovesey, with their emphasis on crime and broken homes as expressions of "negro personality," had been overly pessimistic. "If the authors had left their offices," he lamented, "and gone out into the Negro community, this question, as well as a number of omissions and silly assertions, might have been avoided."[108] Whether or not Black self-esteem was as low to begin with as psychologists presumed is likewise difficult to determine.

The more positive assessments of Black self-esteem that emerged in the late 1960s and early 1970s are hard to interpret as they were published in a climate of fear in which they might accidentally be labeled racist. This encouraged researchers to emphasize positives. Indeed, Black activists had urged social scientists to do this. Studies now reached rosier conclusions. In 1971, (white) sociologists Morris Rosenberg and Robert G. Simmons published their landmark study, *Black and White Self-Esteem*. Like Keil, they criticized earlier self-esteem studies for their sweeping conclusions. In Spring 1968, Rosenberg and Simmons examined self-esteem among a large, representative control group of Black and white children in Baltimore. They concluded, much to their surprise, that Black and white self-esteem levels were about the same. According to the most basic precepts of self-esteem theory, this should not have been the case. In principle, African Americans should have shared white people's negative evaluations if self-regard truly did depend upon others' opinions and attitudes. Yet, in Baltimore, Black children were withstanding the constant onslaught of white racism, a conundrum which the authors could not explain, describing it as a "puzzle." Possibilities opened up that children of color might somehow retain self-esteem even when facing prejudice and discrimination.[109]

Of course, the situation might well have been very different for children living in the South. But we might also take into consideration possibilities that the researchers' findings simply failed to match their preconceptions about African American wellbeing. That said, one problem with orienting toward positive renderings of Black wellbeing was the potential to overlook, even deny, very real emotional issues resulting from ongoing racism, segregation, and discrimination. The danger existed of transmitting a false message: that all was now well in Black communities. Nothing further to worry about.[110] This was definitely not the case. But even Black displays of pride, respect, and self-esteem themselves risked giving off this wrongful impression. Once again, the thin line between expressing victimhood and pride becomes visible.

Not that we should entirely overlook possibilities that positive improvements occurred. Early 1970s African Americans, as a group, might have brandished a more revitalized internal emotional makeup than those before the civil rights campaign. Some contemporary studies concluded that the civil rights movement inculcated a sense that "Black is beautiful" and truly encouraged African American children to appreciate their uniqueness, beauty, and special value. In 1971, Baughman wrote: "Times have changed, and the Black's view of himself may indeed be more positive than it was even a few years ago." Baughman thought that earlier researchers had overplayed Black misery and inferiority, and suggested that just because Blacks had learned to *act* subservient or inferior didn't necessarily mean they had actually *felt* that way. They might have resisted the imposition of racialized representation more effectively than white psychologists had presumed knowing full well that their problems stemmed not from unconscious desires to be white, but in the political and social repression upheld by white society.[111]

Others reached similar conclusions. In 1972, Baptist minister and political activist Jesse Jackson commented:

> I don't think we really wanted to be just like the white man. We always snickered behind his back. We knew we could dance better, run faster, and cook better. Because we weren't allowed very much, we had to compensate by being superior at a few things. Our ability to find survival techniques testifies amply to our self-love.[112]

If true, Black people had little reason to look inwards to source their emotional discontent. Baughman truly believed that embracing "Black is beautiful" offered more than simple wishful thinking. Black people's views of themselves really did improve. If self-esteem is itself fluid and changeable, it seems plausible that levels can fluctuate over time. But, to complicate matters further still, we should also bear in mind the converse of Baughman's argument: that public expressions of pride, confidence, and esteem, especially when articulated for political purposes, may not necessarily reflect the true inner feelings of the socially disadvantaged.

Evidently, the politics of race were intricate. To some extent, it suited African Americans to concede that their self-esteem had suffered historically due to white discrimination. Politicizing emotional experience in this way confirmed the need to fight racial prejudice. Retaining some elements of the "damaged negro" caricature helped score political points about discrimination as long as it was clarified that psychological damage was being consigned to the past. One useful approach that bridged these contradictory perspectives involved looking back in time to confirm historical injustice, while making clear that African Americans had now re-evaluated themselves and began to build self-respect. But was this new-found self-esteem, if it did indeed exist, going to last for long?

The persistence of low self-esteem

In the twenty-first century, self-esteem, self-image, and self-respect problems continue to concern Black communities. In her *Rock My Soul: Black People and Self-Esteem* (2003), bell hooks, argued: "We Black folks have been unwilling to break through our denial and deal with the truth that crippling low self-esteem has reached epidemic proportions in our lives because it seems like it's just not a deep enough diagnosis." She believed that Black people still suffer widely from low self-esteem due to historical experience, white supremacist culture, and self-sabotaging emotions. Offering solutions reminiscent of the 1960s, hooks called upon Blacks to work on their emotional wellbeing and self-respect. In her words: "Living in a white supremacist culture, we as Black people daily receive the message, through both mass media and our interactions with an unenlightened white world, that to be Black is to be inferior, subordinate, and seen as a threat to be subdued or eliminated."[113] Such sentiments would not have seemed out of place in the mid-twentieth century.

In his 2016 book, *Democracy in Black*, Eddie S. Glaude, Jr. wrote extensively about America's "value gap." He believes that American society still values whites more than Blacks, and that everyday racial habits unknowingly sustain this value gap. White supremacy has forsaken extremities such as lynching (although some would argue that "police executions" are not too dissimilar), but subtler manifestations cling on. Black people, every day, still confront the damning belief that white people are more valued than them.[114] Ibram Kendi adds that racist intentions became covert after the 1960s and policies were diluted; although antiracists made some progress, so too did racists and white supremacists.[115] The ongoing concern about self-esteem in Black communities suggests that 1960s activists never truly fulfilled their dreams of an improved collective psychological life, despite their aspirations having been deeply imbued with hope, optimism, and expectations.

Some (usually white) critics suggest that Black violence undermined civil rights progress, making Blacks seem like a resentful, rebellious multitude willing to abuse invitations of equal citizenship kindly offered by white America.[116] However, Kendi speaks instead of the growing alienation of Black people from America's political system since the 1970s, which itself contributed to a resurgence of racism.[117] Civil rights lawyer Michelle Alexander presents this as deliberate political manipulation. Since the 1960s, the American public has been presented with two schools of thought. Conservatives blame poverty and crime on a pathological Black subculture, thereby blaming Black people themselves. In contrast, liberals have called for structural reform to resolve Black crime and poverty. Since the 1970s, some groups have actively denied that race still matters, given the positive developments made in the 1960s, but of course it still does. Alexander writes of a deliberate color-blindness toward ongoing race-related problems, which has been accompanied by a resurgence of white supremacist thought and practice, particularly since the 1980s.[118]

Into the twenty-first century, mass incarceration has become the norm for many Black people. Alexander writes of the isolation, distrust, and alienation caused by mass incarceration, which leads ultimately to self-hatred, in turn creating a vicious circle in which Black youths fulfill society's negative expectations of them. If true, stigma and shame are *still* the common emotional experiences of Black communities.[119] Mary-Frances Winters speaks now of "Black fatigue," caused by centuries of marching, protesting, resisting, writing, orating, praying, legislating, and campaigning for equity and justice.[120] And hooks, to quote her once again, argued: "We Black folks know that our collective wounded self-esteem has not been healed. We know that we are in pain. And it is only through facing the pain that we will be able to make it go away."[121]

It is regrettable that such authors still need to discuss low self-esteem. When hooks spoke to her peers in 2003, the generation of baby boomers for whom 1960s radical politics held such promise, they disclosed feelings of inadequacy, inferiority, shame, guilt, self-loathing, and self-distrust. Like others before her, hooks suspected that low self-esteem remained a major problem for Black people.[122] Together, these contemporary comments on self-esteem are eerily, and tragically, reminiscent of earlier writing on Black self-esteem, before the start of civil rights and Black power movements. Strong parallels exist.

At the time of writing, COVID recently shut down the world. Internationally, riots and protests against George Floyd's (as well as others') deaths have taken place in most major cities. Other than the threat of contagion, race is the burning issue of the day. It seems that 1960s dreams of a better Black future have yet to materialize. What went wrong? In the 1960s, self-esteem was integral to the politics of Black liberation. As with the Human Potential Movement, self-improvement offered promises of a liberated self, in this instance one free from racial discrimination, not so much the emotional ravages of capitalist modernity. The personal and political were intertwined. Black self-confidence was just as visible in popular culture as it was in academic research and the psychological clinic. However, ultimately, the political status quo remained largely unchanged. Discrimination still occurred. African Americans only gradually made their way up the American social scale.

As was the case with counterculture activism, self-esteem initiatives offered the potential for change for only a fleeting period of time in the late 1960s. Nonetheless, and importantly, they reveal the multifaceted nature of self-help techniques, their ability to provide a voice for the marginalized, to express their emotions. Boosting their self-esteem undoubtedly empowered many activists, even if they ultimately failed to achieve sociopolitical change in the long term. The forces of racism, capitalism, modernity, consumerism, social injustice, and socioeconomic inequality all remained in place, regardless of how good people felt about themselves. Nonetheless, as capitalism advanced, particularly from the 1970s, mainstream self-help literature and initiatives cultivated a color-blindness that simply ignored Black people and their specific needs. Overlooking the considerable amount of activist work achieved in the 1960s, personal problems were individualized and isolated from the social realm from where they arose. Self-help, to borrow from Alexander, became color-blind too. How this happened will be explored in the chapters that follow, but first we will explore the often over-looked harnessing of self-help, and the self-esteem concept in particular, among LGBT activists, which also took off particularly from the 1960s.

THREE

Sexuality

Similar to African Americans, lesbian and gay activists had to navigate the paradox of claiming traumatic emotional harm while forging a positive, non-straight identity built upon pride, self-esteem, and self-respect. Here too, inner feelings provided a powerful basis for political discourse. The immediate postwar period was one of enhanced social mobility, although, for many people, this meant social dislocation and isolation. In America, many young gay and lesbian adults who in earlier decades would have stayed in their home towns moved away to larger cities, encouraging sexual permissiveness, experimentation, and freedom to enjoy same-sex relations out of sight of watchful, usually disproving, parental eyes. The relative anonymity of city life made this easier.[1] Nonetheless, LGBTQ communities faced considerable discrimination, which intensified in the 1950s. The homophile and gay rights movements made gains from the 1960s onward, but the 1980s AIDS pandemic later impacted gay communities disproportionately. Against this harrowing backdrop, gay men worked constructively to articulate and, perhaps most challengingly, experience pride at a time when positive feelings weren't always forthcoming.[2]

In reality, public assertions of gay pride often masked ongoing psychological damage caused by stigmatization, homophobia, and disease.[3] "Gay pride," "Gay is good": these eye-catching slogans concealed an intricacy of emotions and feelings. A popular narrative exists that depicts non-straight communities as having traversed an optimistic path toward progress, pride, and social acceptance, although some warn that the requirement to always appear proud dangerously represses, and unofficially bans, negative feelings. As Deborah Gould asks: "Do repeated expressions of gay pride really unmake gay shame?"[4]

The gay liberation period of the 1960s and 1970s was never truly a utopian, radical, hedonistic interlude between the repression of earlier decades and the AIDS epidemic that followed. Shame wasn't simply superseded by pride, although such narratives predominate some accounts of gay liberation, with the 1969 Stonewall riots, sparked by a police raid on a gay club in New York City, highlighted as a crucial turning point. The term "ambivalence" might be more appropriate for recapturing the expansive range of emotions, many of which conflicted, that were experienced by gay and lesbian communities.[5] In 1972, writer Merle Dale Miller made this notably ambivalent statement: "Gay is good. Gay is proud. Well, yes, I suppose. If I had been given a choice (but who is?), I would prefer to have been straight. But then, would I rather not have been me?"[6] Feelings such as loneliness, isolation, despair, and shame are rarely foregrounded in queer history, clashing as they do with a preferred history of positive collective action.[7] The humanistic, hedonistic exploration of the 1960s and 1970s succumbed to a colder 1980s of hostile Republican policies, neoliberalism, and the AIDS crisis. Gay communities, so defiant in their newfound freedom in the 1970s, broadly adopted the new, more sober ethos.[8]

In no other Western country was homosexuality penalized as much as it was in mid-twentieth-century America.[9] Much of this chapter focuses on California, a region long associated with lesbian and gay life, partly due to San Francisco's history of sexual permissiveness. The city hosted a disproportionately large non-straight community and, as a result, left behind a rich historical source base compared to many other regions. Of course, Californian experiences by no means represent the entirety of American lesbian and gay experience, and provide far less insight into the lives of those residing in less accepting regions and contexts. It should also be noted that areas of California, particularly the Orange County, were home to postwar reactionary conservatism.[10] Nonetheless, sources such as LGBT magazines, despite their problems and limitations, are still useful for unearthing historical emotional experience, despite mostly disclosing information on people more comfortable with "coming out" and adopting an openly public non-straight persona.[11]

Sickness and psychopathologies

After World War II, writers on homosexuality discussed same-sex relations with increasing reference to psychological ideas, in some ways mirroring developments in research on race. Psychoanalyst Edmund

Bergler influentially developed a psychotherapeutic interpretation of homosexuality, which he described as "an unfavorable unconscious solution of a conflict that faces every child."[12] Psychologists and psychiatrists began scrutinizing the interpersonal environments of families deemed responsible for "turning" some children gay, usually blaming the boy's relationship with his mother. In his 1967 *The Making of a Homosexual*, Roger Blake (a pseudonym for John Felix Trimble, a critical Christian author on sexuality) drew from these ideas to describe deficient mothering as "the principal contributory factor in the making of the homosexual youth," although he added that "the father too, or simply lack of a father, or a weak and ineffective father, can play an important part." In such accounts, homosexuality was a symptom of an underlying psychological defect.[13]

Many mental health professionals believed that gay men would rather be straight, and they developed "therapeutic" techniques to help them achieve this, the so-called "medical model" of homosexuality.[14] In his 1968 book, *The Overt Homosexual*, psychiatrist and psychoanalyst Charles W. Socarides, a prominent opponent of homosexuality, asked: "Do homosexuals want 'another chance'? I believe that even in those who consciously disavow any such wish and who adamantly defend the nature of their relationship, there exists a deep unconscious desire to alter what early environment has so cruelly forced upon them."[15] These perspectives gained much influence during the war. Before then, neither the army nor the navy had been overly concerned about sexual orientation when screening men for military service. However, in October 1940, more than 16 million men registered for the draft. Due to this large number, the armed forces were in a position to discourage certain groups or individuals from enlisting, including women, African Americans, and homosexuals, the latter seen as a threat to morale, decency, and discipline. By this time, psychiatrists and psychologists had gained much authority in the armed forces. They developed new screening procedures to discover and disqualify homosexual men, and moved their attention from homosexual acts toward spotting and understanding the homosexual himself.[16]

Once war ended, American society sought stability. In practice, this involved upholding the importance of domesticity, the nuclear family, and heteronormativity. In the 1950s, the average marriage age dropped drastically to the late teens and early twenties. The domestic housewife became a key icon, a potent symbol of domestic and national security, if perhaps an imaginary one.[17] The homosexual was her threatening polar opposite. Fear of communism intersected with apprehension about gender and sexuality, triggering a "lavender scare" involving the political targeting of homosexuals. Gay men lost their jobs en masse. Public concern was limited. Doctors cast homosexual men as perverts, moral weaklings and "sexual psychopaths."[18] Some doctors questioned the medical pedigree of such claims, and while the medical model undoubtedly stigmatized homosexuality, it sometimes softened the harsh punitive blow of the criminal justice system. A language of illness, rather than immorality, could invoke sympathy and compassion, and many judges considered the "vice patrol" approach excessive.[19] Nonetheless, the general prevailing atmosphere was one of hostility toward gay men and women.

During the war, psychologist Robert Allan Harper served on the War Manpower Commission. He continued working as a military psychiatric social worker until 1946 before securing a position at Ohio State University, while continuing his private psychotherapy work. During his career, he published numerous books on psychotherapies and "rational living." Harper rose up the ranks of the American Psychological Association. His 1963 article, "Psychological Aspects of Homosexuality," exemplifies the "medical model" approach. He admitted that his article generalized about male homosexual life based on a skewed sample of individuals who willingly sought out psychotherapeutic assessment. In his clinic, Harper probably rarely, if ever, encountered the "socially functional homosexual." Nevertheless, he surmised that "non-acceptance" of heterosexual feelings triggered low self-esteem, deep-seated feelings of inadequacy and insecurity, and an exclusive, "compulsive" adherence to homosexual pleasure. He thought that confining oneself to one "outlet" of orgiastic satisfaction, homosexual intercourse, was "neurotic and perverted." It was not the homosexual act itself that Harper found problematic, but that these weren't accompanied by regular intercourse with women. He argued that although heterosexuals similarly enjoyed intercourse with one sex only, the opposite one, unlike gay men, they had society's approval, and thereby evaded psychological disturbance.[20]

Harper portrayed homosexuality as one of various outlets to which the "emotionally disturbed" turned, stating that "low self-respect and lack of genuine self-love are characteristic of emotionally disturbed people generally." Those who chose homosexual behavior coupled their "anti-sexuality" feelings with low self-esteem; others turned to schizophrenia or addiction. The determining factor was the channel of self-expression that was either encouraged or blocked during the person's child development stage.[21] Same-sex intercourse was a symptom of psychological distress; homosexual behavior was illness. Harper portrayed men who engaged regularly in homosexual intercourse as afflicted by feelings of insecurity, inferiority, and immaturity, which propelled them toward sexual deviance. In the short term, fulfilling homoerotic desire assuaged negative inner feelings, but emotional lives underpinned with such an unsteady psychological foundation crumbled easily when serious life problems arose, at which point, Harper believed, negative self-feelings overwhelmed homosexual men who only then sought psychotherapeutic help.[22]

If the deviant's sex life was indeed linked to low self-esteem, an inferiority complex, and emotional immaturity, as Harper believed, then a logical solution presented itself: increase both self-esteem and heterosexual behavior. He wrote:

> The patient who develops increased self-esteem and less general anti-sexuality, and is no longer addicted compulsively to his former habits of thinking, feeling, and acting exclusively homosexually, becomes able to make more rational and realistic evaluations of his sexual activities. And viewed realistically and rationally, there are certainly a great many more fully satisfying, self-fulfilling, low-risk opportunities for heterosexual than for homosexual functioning in present-day society. The individual who has been released from his anti-sexual attitudes, his cripplingly low self-esteem, and his slavery to deeply ingrained habits, will take advantage of these satisfying, self-fulfilling, low risk opportunities. No longer as neurotic, no longer as controlled by irrational habits, he will relate more realistically to the sexual climate of his society.[23]

Accordingly, in his therapy, Harper sought to convince his patients that exclusive adherence to homosexual intercourse was their main problem:

> They [homosexuals] are really puritanical non-realists and anti-rationalists, self-haters, victims of strong habit-conditioning and frightened, ritualistic avoidance of heterosexuality. This is how they are sick. How can they get well? Not by trying to drive "the beast of homosexuality" out of the systems. On the contrary, they should love and pet and accept this "beast" as the one currently desirable expression of their healthy and human search for sex activity. But while nurturing and carrying this fine animal, how about opening the door to another splendid household pet: namely, heterosexuality?[24]

Harper impressed upon his patients the desirability of *also* developing heterosexual tendencies despite noting an "obstinate and fearful refusal to make a consistent effort to develop new behavior patterns." If "heterosexual success" was achieved, self-esteem would naturally increase, the implication being that such patients would "see the light" and switch to exclusive, permanent heterosexuality.[25]

The consequences of such ideas? Across the West, psychologists developed behavioral techniques intended to turn gay men straight. These involved showing their patients erotic homosexual images of men, and then applying electric shocks to the feet, or inducing vomiting to nurture negative associations with same-sex intercourse. Dating women was then encouraged.[26] The underlying premise was that homosexuality was sickness, but the "cures" only worsened negative feelings toward oneself.

Respectable gays

After World War II, "homophile" organizations emerged that represented the respectable face of the burgeoning gay rights movement.[27] Because of their efforts at demonstrating gay respectability, they are often remembered disparagingly as respectable suited white men debating at committees, or as conservative killjoys frowning upon gay sexual excesses.[28] Seeking civil rights and social acceptance, homophiles demanded integration into mainstream institutions. Mattachine Society and Daughters of Bilitis were the two of the main groups representing gay men and lesbian women respectively. Through publications, and support from some ministers, lawyers, and scientists, they interjected into mainstream society a new discourse about homosexuality that promoted images of homosexuals as socially responsible and productive individuals, perhaps not too dissimilar from their straight counterparts.[29] These groups actively opposed the "medical model," altogether refuting suggestions that they were sick people. Nonetheless, the homophile movement distanced itself from flamboyant aspects of gay life: drag shows, sexual excesses, and so on.[30] Co-founder of the Mattachine Society, Harry Hay had spent most of his childhood and adolescence in California. From the 1930s, he was a communist activist, but he struggled to have his sexual preference accepted by party members. At that point, Stalin was persecuting homosexuals with a vengeance. Only from 1950 did Hay broach the subject of a homosexual rights organization. The Mattachine Society was modeled on the Communist Party, with its secrecy, cells, hierarchical structures, and centralized leadership.[31]

Sociologist and criminologist Edward Sagarin, the so-called "father of the homophile movement," writing under his pseudonym Donald Webster Cory, published *The Homosexual in America* in 1951, and *The Homosexual Outlook* in 1953, books that urged gay men to positively accept themselves and forge collective solidarity.[32] Cory hoped that writing sympathetically from a homosexual, not clinical, perspective, would endear American society to gay men. These were very different books than those being written by psychologists. They demanded that the voices of homosexual men be listened to, not just heard through the observations of doctors, juries, and churchmen.[33] But what experiences became visible? Cory wrote that:

> It is not only shame at my own debasement that demoralizes me, but a great wave of self-doubt that is infinitely more difficult to cope with. Am I genuinely as "good" as the next fellow? Is my moral standard as high? Are my ethics, before my own self and my own Maker, as defensible as those of others around me? Or am I actually what they call me?[34]

Cory added that "the wave of self-doubt that I must harbor" was one of the greatest repercussions of harmful social attitudes. He continued:

> I carry around with me a fear and a shame, I find that I endanger my confidence in myself and in my way of living, and that this confidence is required

for the enjoyment of life. In fact, I must ask myself whether the life that I lead is an inferior one, and whether those who practice it are inferior people.

Ultimately, Cory concluded: "I find it difficult to reconcile self-pride with cowardice, abnegation, the wearing of the mask and the espousal of hypocrisy." The solution was not therapeutic intervention but, instead, self-help and self-acceptance. To achieve this, Cory advised his readers to "reject the entire theory of the inferiority status which the heterosexual world has imposed upon us."[35]

In this way, Cory, like the homophile movement in general, shifted the finger of blame for inner turmoil and psychological distress toward discriminatory social environments, turning on their head arguments that homosexual problems stemmed from inner psychopathologies. Rather, homosexuality was natural, even healthy. Cross-species studies seemed to confirm that same-sex relations were widespread across the animal kingdom, and also in many non-Western human cultures.[36] In 1948 and 1953, the United States was rocked by the publication of *Sexual Behavior in the Human Male* and *Sexual Behavior in the Human Female*, known popularly as the Kinsey Reports. Alfred Kinsey's reports provided unexpected data on the nature of American sexual activity. Shockingly, it revealed that the majority of the nation's citizens regularly violated moral standards and even state and federal laws.[37] One of Kinsey's most shocking revelations was that nearly 40 percent of his male subjects had experienced at least one homosexual act to the point of orgasm.[38] Kinsey offered firm evidence that same-sex activity was surprisingly (for some, alarmingly) common across America, a finding that played a significant role in reshaping ideas, but also concern, about homosexuality.[39]

Psychiatrist and criminologist Donald J. West, writing in 1955, observed that "it is now recognized that homosexuality is an extremely common condition and the only reason why psychiatrists and law courts are not completely swamped with cases is that the great majority of those affected neither seek psychological advice nor fall into the hands of the police."[40] West implied that homosexuality was common and, in light of that, perfectly natural. In turn, this allowed homophile groups to encourage gay and lesbian communities to consider their sexual feelings and urges as natural, healthy, and moral.

These new approaches were expressed in a burgeoning gay press which connected gay men nationally and internationally and provided advice on legal matters and sexual health.[41] These periodicals had a limited circulation, but were nonetheless influential. As well as covering gay life, they reviewed relevant publications and debunked damaging psychological theories. Importantly, they were written by homosexuals themselves, offering new ways of understanding homosexuality beyond the unfavorable assessments of psychologists and social scientists.[42] For instance, a 1963 article in the *Mattachine* commented that early sexual experiences with masturbation or with boys "behind the barn" triggered feelings of guilt, shame, and degradation. As boys grew into men, they faced bleak social messages about "queers" and laws restricting same-sex activities. Both reinforced feelings of low self-worth. Such articulations of emotional experience were intended to normalize inner feelings which many gay men had experienced in confusing, disorientating ways.

The Society for Individual Rights (SIR) recognized the need for gay men and women to create, and participate in, a strong united community. In its first edition, SIR's magazine, *Vector*, issued a policy statement that outlined how the group wanted to give the (homosexual) individual a sense of dignity before himself and within his society. Self-respect was central to SIR's approach, a first step toward ending unfair job dismissals, police harassment, and state interference in personal sex lives.[43] In 1965, a sympathetic Californian, the Reverend A. Cecil Williams, wrote: "To get social justice, we must start by asking who we are. What does it mean to be a homosexual? How can the homosexual best realize his self-potential as a human being?" Williams believed it was possible to be both gay and proud of oneself. It was possible to obtain a sense of worth and dignity even when facing social hostility, partly by creating common bonds and interests with others in similar situations.[44] Both groups refuted sickness models of homosexuality. Not interested in homosexuality's etiology, they inquired instead into how gay people actually lived.[45] Psychologists usually depicted gay men as incapable of lasting love or relationships, feeding stereotypes of the unhappy, unstable homosexual. As an alternative, *Vector* portrayed gay men as perfectly capable of forming healthy, happy relationships with one another, similar to straight couples. The North American Conference of Homophile Organizations (NACHO) was a collection of homophile organizations that met annually in the late

1960s. At its 1968 Chicago conference, activists agreed upon the powerful slogan "Gay is Good," modelled on "Black is Beautiful." The adoption of the word "gay" signaled a rejection of accommodationist approaches.[46]

However, the homophile movement has been accused of unintentional homophobia. It wasn't particularly revolutionary or radical, but simply wanted integration and equal citizenship.[47] The Mattachine Society disapproved of any behavior that might undermine ideas that gay men were respectable and able to enjoy a monogamous family life. Striving for decency, in 1965, *Vector* warned that feelings of worthlessness encouraged men to become promiscuous and risk infection from STDs, adopting a moralizing tone toward homosexual behavior.[48] *Vector* was similarly wary of fleeting sexual encounters. In 1966, it criticized gay men who thought that "not having the burden of children and a nagging wife is a blast," insinuating that gay men should aspire to join a straight world of family life and long-term relationships: "If the homophile philosophy can offer nothing but this bit of homespun sophistry, we are in for a rather unprofitable struggle with the reactionary sections of our society."[49] *Vector* advocated standing up for gay rights democratically and with "quiet dignity and self-respect," hoping that, "as our image of ourselves improves, so will the heterosexual community's."[50] Activists and historians have looked back unfavorably on the Mattachine Society's moralizing tendencies although, in fairness, the group did cautiously pave the way for later developments.[51]

Despite this push for gay pride, public attitudes softened only very slowly. In the 1960s, a number of journalistic articles on homosexuality were widely discussed, some sympathetic, others less so.[52] Regardless of their stance, these usually depicted gay emotional life grimly. Articles, documentaries, and even a vinyl album appeared in 1967 that were aimed at non-gay audiences on the "problem" of homosexuality. A purportedly sympathetic *Look* article, entitled "The Sad Gay Life," drew from research into gay men and quoted a doctor who stated: "To speak of a healthy, happy homosexual is a euphemism, similar to speaking of a cripple or partially blind person as being happy." *Vector* was furious.[53] In a follow-up article, vehemently entitled "The 'Sick' Unhappy Homosexual ... Time to Get Well," *Vector* commented that the media chose not to interview doctors who disagreed with the medical model of homosexuality, accumulating a very selective, skewed expert testimony: "For their own mental health and wellbeing, they [homosexual men] must throw off the cloak of guilt in which they have permitted themselves to be shrouded."[54] In 1971, Merle Dale Miller tried to normalize "gay misery" by pointing out that "none of my homosexual friends are any too happy, but then very few of my heterosexual friends – supposed friends, I should say – are exactly joyous either." Miller added that he was not promiscuous, and knew same-sex couples who had been together for decades.[55]

Established in August 1970 in California, *Gay Sunshine* was a tabloid that reported on the gay liberation movement as it gained momentum, representing a more militant breed of activism than the respectable homophile movement.[56] In its earliest editions, contributors debated, reaching no firm conclusion, whether acting camp or feminine articulated pride or self-hate. Brian Chavez insisted that campness was a positive aspect of gay identity. Refusing to act "camp" signaled self-hate, so central was it to gay identity. In contrast, Allen Young considered campness to be evidence of self-hate as, like white, straight America, it feminized homosexual men and confirmed their status below women, perpetuating and sustaining discriminatory perceptions.[57]

Other *Gay Sunshine* contributors openly criticized homosexual men who treated each other disrespectfully. Some feared that the growing availability of male bodies and promiscuous sex in San Francisco's open culture distracted potential activists from the radical goal of gay emancipation. In October 1970, activist Morgan Pinney contributed an article about being regularly stood up by men, which he believed "displays a large amount of self-hate on the part of those of us who act in such a manner, a great lack of respect for the gay relationship." Not only incensed by rudeness in dating circles, Pinney warned that if gay men held each other in such low regard, society would hardly view them any more positively:

> [This] serves to continue the humanly destructive myths that gay sex is only for quick tricks, that another man is but a cock to suck and then discard like a banana peel once the immediate pleasure is extracted. It strikes me that such frivolous attitudes only reinforce our own self-hate about our form of sexuality which the system has taught us. And all the political awareness and getting it together rap session are absurd among a group of people who in reality hate themselves. If "gay is good" we should act more as if it were.[58]

Inspired by Black Power, gay activists weaponized psychological self-help ideas, including self-esteem. In 1968, Los Angeles-based homophile magazine *Tangents* commented that homosexuals, like members of other marginalized groups, suffered from "diminished self-esteem, doubts and uncertainties as to their personal worth, and from a pervasive and unwarranted sense of their homosexual condition." *Tangents* advocated a new "positive, affirmative attitude toward themselves and their homosexuality" to replace this "wishy-washy negativism" and establish in homosexual communities "feelings of pride, self-esteem, and self confidence in being the homosexuals they are and have a moral right to be."[59] In 1969, *Tangents* elaborated further on the positive impacts of adopting the "Gay is Good" lifestyle: "You gain self-respect, self-esteem. People don't respect someone who doesn't respect himself. If you respect yourself, people will respect you. If a person is homosexual and respects himself as a homosexual, then the public is going to do the same."[60]

From the 1970s, gay activists, shedding their aura of respectability, aggressively criticized the so-called "shrink pigs": those psychiatrists and psychologists who "speak with a confidence and authority that would be laughable if it were not for the infinite suffering they cause."[61] While favorable toward Cory's *The Homosexual in America*, the main targets included books by Edmund Bergler, Irving Bieber, Albert Ellis, Charles Socarides, Lionel Ovesey, and Lawrence J. Hatterer.[62] Self-help, which incorporated self-esteem, remained important to 1970s gay activism. In 1979, Gay Liberation Front activist Leo Louis Martello penned an article in *Gay* magazine which argued that everyone needed a strong sense of self-esteem. All gay people should work on their self-esteem for "no matter how oppressed any minority, how unjust the system, how unfair the discrimination, each person is responsible for his own life ... the shiftless, goalless, lazy gay person cannot be a spokesman for gay power." Martello equated gay power with "economic security, self-esteem and a sense of achievement."[63] In 1979, *Update*, another gay magazine, gave away a free questionnaire tool for measuring self-esteem.[64]

Evidently, the burgeoning late 1960s gay rights movement drew influence from contemporaneous Black strategies that encouraged marginalized individuals to work collectively. Undoubtedly, campaigners were overly optimistic in their hopes that, simply by altering mindsets and self-perceptions, homophobia would melt away. Nonetheless, the trope of self-esteem offered a useful symbolic weapon for confronting a hostile American society, providing a basis for pride, respect, and social recognition.[65]

Lesbian self-esteem

Lesbian life also changed. The large-scale entry of women into the workforce, geographical mobility, and a new concentration of women in the armed forces created opportunities for lesbian women to meet.[66] Many lesbian women suspected that gay male activists considered their emancipation to be secondary and less important.[67] In the 1970s, a number of magazines emerged catering specifically to lesbian, and sometimes female bisexual, needs and seeking to reclaim ownership of female bodies and sexualities from men – even from gay men. Their activities intersected with second-wave feminism's openness about speaking of sexual matters. The magazine *Lesbian Connection*, founded in 1974, wrote: "Why masturbate? Masturbating is a way as women that we learn eroticism. We learn to respond to ourselves sexually, we learn to build self-esteem, and most important we become sexually independent. You do not have to ask someone to make love to you."[68]

Amazon Quarterly, a self-styled lesbian feminist arts journal, regularly published interviews and short autobiographical accounts of lesbian experiences that assured readers they were not alone in navigating socially unaccepted sexual feelings, and that same-sex feelings and urges were perfectly normal. Not all accounts dwelt on the emotional misery of sexual confusion, and those that did usually concluded with a happier ending. However, interviewees did disclose negative initial experiences. Their accounts provided an intimate, less politically charged rendering of lesbian inner life than the public-facing pronouncements of male gay activists.

In 1972, one contributor, Flo, wrote about feeling oppressed by gay men. "It's just the way they trivialize everything I do," she exclaimed, adding that "gay men frequently criticize and put down women." Flo remembered first feeling strong affection toward another girl in her early teens. Although she didn't think much of it at the time, Flo lamented: "Then, when I read about, when I found the books about homosexuality in psychology, then I thought something was wrong with me." The books in question were a medical dictionary and a Freudian text, which caused Flo considerable anxiety. As she recollected, confronted with medical evidence claiming she was psychologically ill, "I didn't know if they were right or not. I didn't think they were but I wasn't sure." It was only when she met her long-term partner Jeanne at the age of 20 that Flo felt more confident about the normalcy of her sexuality.[69]

Another interviewee, Laura, commented on the mental uplift that occurred when she accepted herself as lesbian, happily stating:

> Oh, it's an unbelievable improvement. Everything that's happening is me, purely for myself, and it's wonderful. It's the best thing I ever did because it used to be such self-hatred for me. My feelings about women were so heavy that when I did come out it was completely and I just left all those regrets. Now I feel that I couldn't be anything worse in a way, than being a lesbian, that was my ultimate self-hatred, so everything I do now is okay.[70]

Such accounts offered redemptive stories that encouraged lesbian and bisexual women to embrace their feelings for other women independently of gay male culture. Notably, these were less concerned with sexual acts than male homosexual accounts of the time, and also foregrounded the personal over collective action (although the latter theme still figured prominently in contemporary lesbian literature). However, emotional wellbeing was not always easy to secure. Activists observed that suicide was far too common among lesbian women.

Del Martin and Phyllis Lyon, a couple, were well known in feminist and gay rights activist circles.[71] In 1955, they co-founded the Daughters of Bilitis, the first American social and political organization for lesbians. They remained active in San Francisco for many decades, before finally marrying in 2004. Martin and Lyon considered lesbian women to have been at high risk of suicidal thoughts when experiencing the identity crisis associated with the realization of both their sexuality and social discrimination. They cited a group discussion held with twenty young lesbian women. At this, it transpired that eighteen had attempted to end their own lives. These women "had been made to feel so degraded by the realization of their lesbian identity, their self-image so debased, that suicide seemed to be their only way out." Martin and Lyon added:

Some never make it through this long and lonely journey. They can't face rejection, the concept of being "queer" or different. They believe the myths and accept what they see on the surface of gay life. They succumb to fear and assume the guilt. They cannot play it straight, nor can they adjust to their homosexuality. So filled are they with self-negation and self-hatred, there is nothing left for them but death.[72]

Being Black and gay

The emotional situation of gay and lesbian communities has been compared to other minority groups eager to replace internalized self-hatred with pride. Lesbian and gay activists considered their plight similar, broadly speaking, to other oppressed groups.[73] Many hoped to connect with African Americans who had similarly weaponized their negative emotional experiences and damaged psychologies. In his sympathetic 1965 book, *Crimes without Victims*, sociologist and criminologist Edwin M. Schur outlined some of these comparisons. Both groups used "passing" (as white, or straight) as a coping mechanism, denying their authentic identity. Both groups had created a social movement, and both had been duped into considering themselves inferior.[74] Nonetheless, African Americans and lesbian and gay activists regularly bickered over who was most oppressed. Lesbian and gay activists highlighted their lack of basic social rights, such as marriage. Black people pointed out their historical enslavement, centuries of oppression, and inability to disguise skin color.[75] African Americans argued that non-straight people weren't socially segregated; many were well educated, socially mobile, and financially comfortable. A vocal minority were furious that homosexual groups had applied Black models of oppression and victimhood to advance their own cause. As one person argued: "To be gay is merely an inconvenience, to be Black is to inherit a legacy of hardship and equity." Evidently, the issue of victimhood was up for dispute.[76]

Important too were the intersecting layers of discrimination facing Black gay men. Homophobia, as with sexism, was rife in Black communities, even among activists challenging racial discrimination. Gay African Americans endured discrimination from both society and their own communities. To worsen matters, as Merle Dale Miller observed in 1971, "homosexuals, unlike the Blacks, will not benefit from any guilt feelings on the part of liberals. So far as I can make out, there simply aren't such feelings."[77] In the late 1960s, the ever acerbic Eldridge Cleaver controversially wrote: "I, for one, do not think homosexuality is the latest advance over heterosexuality on the scale of human evolution. Homosexuality is a sickness, just as are baby rape or wanting to become the head of General Motors [a car manufacturing company notorious for shutting down factories staffed largely by Black labor]." Cleaver also took a swipe at James Baldwin's homosexuality, claiming that such Black men were "acquiescing in their racial death-wish." Attempting a sophisticated psychotherapeutic explanation, Cleaver explained that Black men turned to homosexuality because of a deep-rooted frustration at being unable to give birth to a white man's baby. In his words, "The cross they have to bear is that, already bending over and touching their toes for the white man, the fruit of their miscegenation is not the little half-white offspring of their dreams but an increase in the unwinding of their nerves." Elaborating further, and developing further dubious psychotherapeutic ideas, Cleaver wrote that Black homosexual men had been psychologically castrated and stripped of their masculinity by white society. Some even took white men as their lovers, a symptom of their inner self-hatred of Blackness.[78]

Being Black and gay meant facing prejudice on the grounds of both racial and sexual identity.[79] Some believed masculinity to be part of the problem. Writing in 2000, Earl Ofari Hutchinson reflected on why he had harbored homophobic attitudes, especially toward Black gay men. He recalled that:

> From cradle to grave, much of America drilled into Black men the thought that they are less than men. This made many Black men believe and accept the gender propaganda that the only real men in American society were white men. In a vain attempt to recapture their denied masculinity, many Black men mirrored America's traditional fear and hatred of homosexuality. They swallowed whole the phony and perverse John Wayne definition of manhood, that real men talked and acted tough, shed no tears, and never showed their emotions.

Hutchinson added: "Black gay men became the pariahs among pariahs, and wherever possible, every attempt was made to drum them out of Black life. Even James Baldwin didn't escape."[80]

However, there were some positive points of intersection. San Francisco's gay movement found an unlikely ally: the city's Protestant clergy. A. Cecil Williams, the Black minister of Glide Memorial Methodist Church, had long been involved in the Black rights struggle. He encountered many male runaways who turned out to be gay. Some had run away from home due to familial ostracism and sought assistance from Mattachine Society. Gradually, this led to a joint effort to work together for social justice for homosexuals and lesbians.[81] In 1968, Williams delivered a sermon to a 1,000-strong congregation and announced that the church needed to change its views on homosexuality. His sermon, "A New Word and World for the Homosexual," delivered the message: "If Black is beautiful, then gay is good." Williams commented to *Vector*: "I expect to be challenged for this, but it's the only position I can take if I'm to faithfully follow the gospel."[82]

In the following decades, occasional sub-strands of Black gay activism emerged, sometimes with church support. In the mid-1980s, Chicago had a bar, Foster's, which was patronized primarily by Black gay men and the United Affinitas Church, a gay Black congregation. The church held workshops on matters such as AIDS, nutrition, venereal disease, and, above all, bolstering self-esteem. These reportedly drew up to sixty attendees each week, the chief message, according to Max Smith, the head of the church board, being that "it's right to feel good about loving someone gay." Similarly, assertive groups such as "Black and White Men" picketed together against gay bars that refused to admit Blacks.[83]

Positive gay therapies

From the 1970s, sickness models fell out of fashion, partly due to LGBT activism. Numerous protests were held at American Psychiatric Association meetings and, in 1973, the Association removed homosexuality from its official list of diagnoses, although not with unanimous approval. In theory, homosexuality had been "de-medicalized," but, nonetheless, conversion therapies remained in use.[84] Books such as psychologist Clarence A. Tripp's 1975 *The Homosexual Matrix* decried attempts by mental health professionals to convert gay men to straight, and argued instead for social tolerance.[85] In 1976, Charles Silverstein, director of a New York counseling service called the Institute of Human

Identity, announced that conversion therapy patients were engaged in a sadomasochistic drama. Suffering from depression, low self-esteem, and a masochistic desire to be punished for transgressing proper masculine roles, these self-hating men then took themselves off to therapy. That same year, Gerald Davison, president of the Association for the Advancement of Behavior Therapy, called for a complete halt to behavior modification.[86]

In response to the hostile (if improving) attitudes of mental health professionals, gay counseling centers offering less punitive therapy opened in some cities, including New York, Philadelphia, Pittsburgh, Boston, Seattle, Minneapolis, and San Francisco. These provided a low-cost service to gay people experiencing emotional distress but who had no wish to change their sexual orientation. Rather than obsess over "cure," gay counseling centers affirmed homosexuality as normal and provided positive approaches.[87] The Human Potential Movement and humanistic psychologists were strong influences given their rejection of professional tendencies to dwell on psychopathologies.[88]

In 1979, psychotherapists Betty Berzon and Robert Leighton published *Positively Gay*, a collection that outlined the trauma involved in accepting oneself as gay but also the psychological joys of coming out.[89] Meanwhile, popular gay magazines such as *Advocate* urged therapists to adopt approaches oriented toward self-acceptance and self-esteem.[90] Radical gay therapists led the way. In the 1980s, Rob Eichberg facilitated Experience Weekend workshops across America. These drew extensively from the Human Potential Movement and New Age ideas. Eichberg wanted to transition his clients from depression and powerlessness to a powerful and empowering state of consciousness. Victimhood meant a lack of self-respect, self-esteem, self-acceptance, and self-love, so Eichberg told his clients, adding that injustice wasn't something that simply happened to them. Their own depression and lack of self-esteem created and perpetuated victimization.[91]

Of course, one might argue that such approaches transferred responsibility for the emotional ramifications of discrimination back on to individuals. An additional risk was the potential to overlook very real psychological problems while producing optimistic renderings of lesbian and gay psychologies. As discussed earlier, it became unfashionable in the 1970s to pathologize African American life, inadvertently or otherwise, and it was politically safer to argue that Black/white self-esteem was roughly the same. However, this risked denying racism's ongoing emotional effects. Research into homosexuality similarly abandoned pathologization to advance rosier assessments of lesbian and gay emotional wellbeing. However, there were potential pitfalls in a new generation of researchers applying "no-difference/no-pathology/no-problem" models to other marginalized groups and potentially sending false messages stating that all was now well in the minds of lesbian and gay communities, despite homophobia's lingering presence.[92]

Despite these problems, the general trend undoubtedly shifted toward understanding homosexual life. In 1956, psychologist Evelyn Hooker published a series of articles on Los Angeles gay life. Hooker had jettisoned pathologization to explore instead how gay men lived and experienced their world. This was a more ethnographic approach.[93] This trend accelerated in the more buoyant gay pride atmosphere of the 1970s. In that decade, William E. Cross, Jr., and later Thomas Parham and Janet Helms, developed the Racial Identity Attitude Scale (RIAS). Psychologists adapted this in order to comprehend how identities (positive and negative) formed in lesbian and gay communities, presuming that broadly parallel processes existed across socially marginalized groups. While the primary focus was on gay men, a smaller number of initiatives and books foregrounded lesbian and bisexual women.[94]

Don Clark was among the first openly gay psychologists. Before coming out of the closet, Clark had fathered two children. Over the years, he came to terms with his sexuality and, as a San Francisco-based therapist, he set out to help gay patients do the same. Clark formed the Association of Gay Psychologists. In 1977, he managed to publish *Loving Someone Gay*, despite facing a formidable number of rejections from leading publishing companies. Later, the book was republished in various editions and languages. In it, Clark spoke out about the subtle effects of discrimination and the creation of a "damaged self-concept," drawing openly from similar appeals to the emotions made in African American campaigns. He complained about a lack of homosexual visibility in the media, the inability to marry, and barriers to secure employment. If gay people became more visible, emulating contemporaneous African American strategies, straight America might rethink its negative values and attitudes.[95]

Clark believed that, in some ways, gay men experienced worse discrimination than (straight) African Americans, for "when the wounds are inflicted by family, friends, trusted counselors, and civic leaders, the wounds go deep and damage self-esteem. A scarred and seriously damaged self-concept is the result." Clark relayed this as follows:

> The gay person begins to think of himself or herself as wrong, bad, or defective. The one hope is that no one knows, no one sees. The secret must be protected, sometimes with lies that reinforce the bad self-evaluation. "Not only am I sexually perverted," one reasons, "but I am also a liar." Energy goes into a personal civil war fought against natural gay feelings. Privately, we grow up during the invisible years suspecting that there must be many basic things wrong about us. Why else would loved ones say such things about people who share our feelings? And the seeds of self-doubt and self-hate grow and grow. They live with us every day. Only the lucky few, so far, have found ways to excise them.[96]

Clark depicted an internal war against unwelcome sexual urges and feelings that worsened as these inevitably transmuted into physical, sexual behavior. As homosexual experiences accumulated, self-hatred increased. For that reason, too many gay men committed suicide. Self-esteem could be irrevocably deflated. Even for those who avoided suicide, "the surviving wounded gay must cope with anguish and despair." Some settled for a zombie's life of work, food, television, and sleep; others sought institutional help; still others tried to numb their pain with drugs and alcohol. All of this meant stepping back into the closet rather than striving for self-respect.[97]

To avoid all of this, Clark organized weekend groups on growth, an alternative to "conformity or self-destruction." Drawing from ideas about self-actualization, these raised the consciousness of gay men and aided their growth into full human beings. Among Clark's arguments was: "Loving someone gay will liberate you from irrational fear, from dangerous misconception, from crippling inhibition."[98] One can sense here the ongoing struggle between painting too gloomy, even pathogenic, a picture of lesbian and gay emotional life, or one that is overly optimistic. Ideas of transition and transformation helped navigate this problem, encouraging gay men to accept previous emotional damage while looking forward to a psychologically healthier future.

Similarly, psychoanalyst and psychotherapist Richard A. Isay treated his gay clients as individuals. He had no interest in trying to change their sexual orientation; he addressed their day-to-day problems just as he would his heterosexual clients. Isay maintained that homosexuality was genetically determined, a stance that allowed him to further question the futility of trying to turn gay men straight. He discussed a nexus of interpersonal encounters surrounding young homosexual boys who needed love from both their parents and society to build a positive sense of self-regard and hopefully grow up gay comfortably. This self-assured gay identity would provide a psychological defense against social prejudice. Anger should not overwhelm the psyche, for this led to masochism, depression, and further self-esteem injury. Ultimately, Isay believed that "it is the affirming love of another man that is the most effective antidote to the battered self-esteem."[99] In part, he was reacting against his clinical training which, drawing from Freud, had presented homosexual feelings as a perversion. While at university, Isay convinced himself that his infatuation with another male student was a passing phase, and a stronger interest in girls would soon develop. It never did. Eventually, he sought conversion therapy, and gradually accepted his sexuality.[100]

Counselors, psychologists, and social scientists began to reflect more closely on how people came to terms with, and adopted, their lesbian and gay identities.[101] The term "homophobia" was first used in the late 1960s, and was popularized by heterosexual psychologist George Weinberg in his 1972 book *Society and the Healthy Homosexual*. Weinberg believed that both hetero- and homosexuals could be homophobic, and that both would be emotionally healthier if they freed themselves from the shackles of their prejudice.[102] By the 1980s, increasing numbers of psychologists identified homophobia, not homosexuality, as a leading cause of emotional distress.[103] Developmental models gradually replaced sickness ones, although many still dwelt on discrimination, suicide, depression, and psychological ailments.[104]

"Coming out" began to be framed as a potentially encouraging life phase with enormous self-esteem benefits, but usually only if significant family members, usually parents, accepted their son or daughter's sexuality.[105] Psychologists and counselors saw the period leading up to "coming out" as overflowing with negative emotions about homoerotic feelings: fear, denial, withdrawal. Without assistance, the process of self-identifying as gay might worsen self-hate, fear, and social isolation. As a result, the psychologist's work began to involve helping young men adopt more positive experiences of "coming out," which was itself recast as being emotionally beneficial.[106] It brought opportunities to socialize more openly with other gay and lesbian people and to integrate into supportive networks, with potentially positive mental health impacts. Of course, "self-disclosure," to use the expert term, could lead to social, and even familial, rejection, but psychologists insisted nonetheless that secretly harboring sexual urges and feelings corroded psychological wellbeing far more deeply than rejection. Increasing numbers of people were willing to take the risk.[107]

To better understand the processes leading up to "coming out," many psychologists used an identity formation model developed in 1984 by Vivienne Cass, a researcher who, although she worked largely in a heteronormative climate, approached lesbian and gay life as "normal." Negative social attitudes, not sexuality, were the problem. Gay people were socialized in a society that was heterosexual in outlook, what historian Margot Canaday terms the "straight state."[108] Cass surmised that lesbian and gays typically passed through stages of (1) identity confusion, a burgeoning, but bewildering, awareness of sexuality that lowered self-esteem; (2) suspicions that one might be gay, or at least bisexual, accompanied by feelings of alienation; (3) a phase in which homosexual identity was "tolerated" but not fully accepted; (4) a private identity acceptance, while still "passing" as straight; (5) a period of fierce, identity pride; and, (6) finally, an integration of a non-straight identity into one's self (rather than gayness being *the* identity).[109] However, here once again we see the problem with painting an overly optimistic image. Only from the 1990s did articles start to appear more frequently on the diversity of "coming out" experiences, including negative ones. Lacking social support and familial acceptance, too many LGBT individuals still sank into depression, even suicide.[110]

AIDS

Self-esteem problems were compounded further by a viral epidemic: AIDS. Initially, many doctors and nurses in the early 1980s refused to treat AIDS patients, due to homophobia and fear of infection. Nonetheless,

California led the way in providing humane physical and psychological services for HIV+ men. New mental health agencies included the University of California's AIDS Health Project and California Pacific Medical Center's Operation Concern. The AIDS crisis had a devastating, confusing impact on pride and self-esteem. The campaign to render lesbian and gay life positive continued but, for many of its victims, the AIDS pandemic presented a severe psychological crisis. An HIV+ diagnosis could rekindle negative feelings toward oneself. Deborah Gould identifies a constellation of affects, feelings, and emotions. AIDS discourses were "emotionally saturated" and deeply ambivalent, swinging between fear, shame, pride, grief, despair, fury, and hope. Even before AIDS, the emotional habitus of gay communities was structured by ambivalence; AIDS heightened contradictory feelings. To worsen matters, the medical emergency was initially met with government indifference, since AIDS primarily affected only gay men.[111]

Faced with an imminent, and unpleasant, death, gay men responded in emotionally complex ways. Self-esteem could be entirely depleted. Fears surfaced about the unwanted exposure of sexuality to family and friends. Limited social support was available. Infection could cause self-condemnation. Combined, these reawakened conflict and ambivalence about homosexual identity even among those who had taught themselves self-pride. Having briefly experienced freedom and pride in their sexuality, many men now felt they were being punished, perhaps even by God. Often, an AIDS diagnosis involved an involuntary coming out of the closet to family and friends. At worst, families might reject and abandon their infected sons due to aversion toward their sexuality.[112] Sex therapists were forced to change tactics. One wrote: "I used to encourage gay clients to go out to a bar and pick up people, to try out, explore and experiment with lots of different aspects of their sexuality because I think that's one of the most wonderful, positive aspects of gay subculture. Now with AIDS, I'm needing to do the opposite."[113] The new challenge was to remain sex-positive and gay-affirmative while educating about safe sex.[114] (It is worth noting that little attention was paid to HIV+ women, and not even to the many lesbian and bisexual women who supported gay men through the crisis, often nursing critically ill patients, even those whose families had abandoned them because of their sexuality and/or diagnosis.[115])

The sympathetic San Francisco doctors presented AIDS as a profound psychological, as well as physical, issue, in the face of considerable lack of interest among physicians and psychiatrists. Between 1982 and 1983, San Francisco General Hospital established a multidisciplinary AIDS clinic and inpatient unit, with counseling services provided for both the infected and the "worried well," recognizing the need for emotional support.[116] In his practice, clinical psychologist Walt Odets proudly worked through the crisis with HIV-positive and HIV-negative patients, believing that both suffered in different ways. His view was that all gay men lived with the epidemic, regardless of HIV-status. Survivors suffered extensive loss and grief. In Odets words: "Many men have suffered more loss and survived more death than any one man could be expected to endure." Odets defined himself as a clinical psychologist, a gay man, and a member of San Francisco's gay community. He had personally witnessed the epidemic's terrible toll and had himself lost a number of friends. He described AIDS as a complex sociopsychological experience, a "psychological epidemic" that affected even the uninfected.[117]

Even putting AIDS aside, Odets believed that many gay men grew up with developmental issues that inhibited their daily lives: difficulty expressing feelings, insufficient self-esteem, and ingrained anxiety. Low self-esteem seemed disproportionately higher in gay men than in the general population, largely due to society's treatment of homosexuals. AIDS exacerbated this problem by making being gay synonymous with harboring and spreading a so-called "gay plague." Illness decimated already depleted self-esteem levels. Some projected their inner self-hatred onto gay culture, a manifestation of self-hatred.[118] As one group of researchers put it, infected men "may see themselves as the rest of society sees them, that is, as causal agents of the disease and in so doing they become an object of scorn and self-blame."[119]

Gay self-esteem issues predated the AIDS epidemic, but were undoubtedly deepened during it. A few clinicians feared there could be a massive increase in unprotected sex if hopelessness were to breed feelings of helplessness, passivity, and impulsivity resulting in self-destructive behavior.[120] Terminal illness forced individuals to adopt a new social role, that of a patient, hardly the rejuvenated self-image that gay activists had earlier dreamt of. Gay communities suffered doubly from the stigma associated with disease and sexuality. Diagnosis pushed infected individuals either further into, or more firmly outside, the closet.[121] Odets wrote:

> Homophobia exists within the hearts of gay men themselves ... homosexuals have been considered sick, have homophobically considered themselves sick, and are now sick with AIDS. Homosexuals have always been punished for their sexuality, and may now feel as though they are being punished further with AIDS ... AIDS has allowed a subtle reassignment of familiar, powerful, psychological conflicts to a new issue.[122]

Gay African American men suffered particularly severely. One 1998 research article stated: "The Black church does not embrace its gay children. It shuns them. It has belittled them and made them feel less than human. It adds to the self-hatred, self-loathing you find in many homosexuals." Some felt confused about why God had put these feelings in them and felt certain that they were heading to hell. Several men commented on the difficulty of living as a "double minority." Overall, these lowly feelings encouraged a lack of regard for safe sex among African American gay and bisexual men.[123] When faced with a disease that threatened significant numbers of African Americans, traditional Black community leaders, political organizations, churches, the press, and community leaders refused to respond to the crisis, or, at best, delayed doing anything about it. Mainstream images of AIDS victims tended to be white. Perhaps Black leaders, themselves still striving for social acceptance, were unwilling to associate their communities with a new disease. Historically, Black communities had been blamed unfairly for spreading all manner of diseases including tuberculosis and syphilis, so maybe there existed a reluctance to publicly acknowledge the extent of the problem in Black communities. Homophobia undoubtedly played a role too, as Black gay people still suffered dual marginalization from within their own communities and white society. By 1990, more than 50,000 Black men, women and children had died from AIDS in America.[124]

By the 1990s, more sympathetic journalistic accounts depicted gay men, now more aware of the importance of safer sex, as lacking in self-esteem, and, because of this, as fatalistic.[125] In 1995, the San Francisco AIDS Foundation developed a HIV-prevention campaign on billboards, bus shelters, and television, based around the premise that, "We have to tell them [gay men] they are worth it." Over twenty weeks, the campaign sought to elevate gay men's self-esteem, boost their self-concept, and promote risk-reduction behavior. One poster showed a young African American woman with the accompanying message: "I am worth loving. I believe in who I am. I practice safe sex so I can stay healthy. I wouldn't want to face the physical and emotional ordeal that comes with having AIDS." According to this approach, unsafe sex practices could be resolved simply by making men feel better about themselves.[126] However, in 1995, *Advocate* warned that positive role models of HIV+ men risked glamorizing AIDS. A recent film, *Jeffrey*, had featured an HIV+ "gym god" played by model Thom Collins. The Los Angeles Gay and Lesbian Community Services warned that efforts to remove the stigma of a positive HIV diagnosis risked glossing over the disease's realities and promoting a sense that it was sexy to be HIV+. AIDS had been sanitized.[127]

In some ways, perhaps things haven't improved all that much. In 2005, Alan Downs still described gay emotional life as follows:

> For the majority of gay men who are out of the closet, shame is no longer felt. What was once a feeling has become something deeper and more sinister in our psyches. It is a deeply and rigidly held belief in our own unworthiness for love. We were taught by the experience of shame during those tender and formative years of adolescence that there was something about us that was flawed, in essence unlovable, and that we must go about the business of making ourselves lovable if we are to survive.

Downs added that "as the eye cannot see itself, we cannot see or feel this embedded shame. But make no mistake, the shame is there and it is very real."[128]

As recently as 2019, Odets was still warning that centuries of shame and stigmatization for homosexuals still precludes self-realization. Despite the progress made, LGBT communities continue to internalize social stigma as self-stigma, causing them shame, anxiety, and depression. Calling attention to these emotional problems still risks pathologizing and blaming LGBT communities. In Odets's own words: "Shame breeds self-doubt and fosters an impaired sense of self-worth and competence that encourages unnecessarily limited lives." He added: "The mere assertion of gay pride does not undo it [shame]; it hides it."[129]

Once again, the inherent complexities of using therapeutic culture to improve opportunities for social justice become apparent. It's clear that very specific problems of self-esteem arise within various social groups, and that these have tended not to be captured in either critiques of self-help and therapeutic culture, or in most self-help books that focus on self-esteem. Indeed, self-esteem approaches, rather than being altogether oppressive in nature, have been used to empower, and as a means of expressing negatively felt emotions, even if they do have some potential to pathologize and individualize problems stemming from broader social factors. For gay and lesbian activists, particularly those operating prior to the outbreak of AIDS, the idea of self-esteem was both personally and politically useful, offering them opportunities to engage their emotions with society at large, and within their various approaches to activism and social justice campaigning. Combined with broadly similar approaches being pursued in African American activism, things seemed to be off to a good start. Perhaps self-esteem could really help liberate marginalized communities from the trials and tribulations of the postwar American society. However, this was not to be. Despite all their hard work, the burgeoning self-help and self-esteem movements that developed in the 1970s, and accelerated in popularity from the 1980s, simply ignored all the activities pursued by activists. The following chapters explore the damaging implications.

FOUR

Mothers

In 1985, African American psychologists Anderson J. Franklin and Nancy Boyd-Franklin warned readers that "to deny the existence of racism for the Black parent is to misguide the psychoeducational development of the child."[1] The Franklins knew that Black children faced particular circumstances and experiences, but blamed mainstream psychology for mostly overlooking life-challenging problems such as discrimination. Even despite the attention paid to self-esteem among Black activists, and the inherent value in the concept which they had demonstrated, from the 1970s, most self-help advice simply ignored the existence of minority groups and their vibrant self-help approaches. It addressed the "normal" American, defined predictably as white, middle class, and from a nuclear family, and offered only individual, not collective, opportunities.

It was in the 1970s that the popularity of self-esteem really took hold, around the same time that the counterculture ran out of steam. Some critics lament that the counterculture turned into consumer culture, and human potential was commercialized.[2] The "happiness fantasy," originally intended as a form of protest against capitalist society, was co-opted and integrated into corporate cultures and mainstream psychology. Group empowerment was repackaged as individual advancement.[3] As countercultural dreams faded, the strands of self-help advice more attuned to capitalism became predominant.

Connected to these developments, America witnessed a resurgence of conservative values fueled by a backlash against radical left activism. Feeling threatened by civil, women's, and gay rights, conservatives insisted that traditional values must be restored. They saw American civic commitment and social responsibility being washed away around them. Moral and sexual wellbeing was in grave danger. (Nuclear) family values needed to be reinstated.[4] Despite being female herself, author and activist Phyllis Schlafly was deeply opposed to women going out to work and enjoying equal rights. She saw the family as the fundamental institution and backbone of American society. In the counterculture, it had been attacked, debased, maligned, slandered, and vilified.[5]

A backlash against welfare and civil rights gains helped Richard Nixon's 1969 presidential election campaign, which had seen him play upon the racial fears of white voters. The Nixon administration worked hard to blame social inequality on individual behaviors. Gradually, the politics of civil liberties gave way to one of law and order.[6] As Americans faced difficult economic times in the 1970s and 1980s, many whites used the civil rights movement as a convenient scapegoat for their own economic vulnerability. The politics that emerged subsequently under the Reagan, Bush, and Clinton presidencies rolled back civil rights gains. The politics of self-help was contradictory. Self-help appealed multidimensionally to diverse, even competing, political agendas. For that reason, the idea of self-esteem could support radical agendas of tackling social injustice while also finding a home in conservative politics that advocated individual initiative above collective action. Mainstream self-esteem literature was largely shorn of its transformative potential and its capacity to provide a voice for the socially marginalized, and instead reinforced the individualistic values of capitalism and white American culture.[7]

Robin Diangelo writes that white Americans constantly circulate and reinforce messages that position white people as superior. She describes these as "thousands of stored 'bits'." I consider self-help advice as one of these "bits," and a significant one too. White people are mostly oblivious to day-to-day cultural practices that maintain racism as a network of norms and actions that advantage some but disadvantages others.[8] As Diangelo writes:

> Individualism is a story line that creates, communicates, reproduces and reinforces the concept that each of us is a unique individual and our group memberships, such as race, class or gender are irrelevant to our opportunities. Individualism claims that there are no intrinsic barriers to individual success and that failure is not a consequence of social structures but comes from individual character. According to the ideology of individualism, race is irrelevant.[9]

Certainly, from the 1970s, much self-help advice presented low self-esteem as the *cause* of most people's problems. Indeed, all manner of social problems – crime, alcoholism, child abuse, school failure – began to be viewed (simplistically) as resolvable through simple self-esteem-building techniques, the polar opposite of activist approaches. And if poor self-image was really responsible for all these problems, then the solution – improving one's self-esteem – seemed all the more imperative, even if this did involve blaming life's problems entirely upon individuals. Through simple day-to-day micro processes (reading a self-help book; thinking more positively about oneself), Americans were encouraged to endure, or adapt to, the social situations facing them. They found little in self-help literature to help navigate self-esteem problems caused by race, gender, class, or sexuality-based discrimination. In place of these could be found only oversimplified messages about learning to become happy.[10]

It could be said, from a sociological stance, that the dominant culture (white, straight, male) leveraged the self-esteem concept to its own advantage, using therapeutic culture and positive psychology to advance a certain idealized vision of the "normal," self-confident American. To achieve this, it simply ignored the marginalized and their specific emotional needs. Speaking about race, Imani Perry argues that we now live in an era of post-intentionality, one in which inequality manifests more through unconscious bias than deliberate intent.[11] I agree that erasing social injustice from self-esteem advice probably wasn't part of some purposeful plot, but it did unconsciously articulate messages about the people valued most in American society. Ethnic minorities, LGBTQ+ individuals, even women and girls for a while, rarely found much of relevance to their specific needs in the self-help literature that first became so popular from the 1970s, and which seemingly contained universal "truths" about inner wellbeing applicable to everyone. Regardless of these basic issues, self-esteem advanced into one of psychology's (and self-help's) core concepts.

As the Franklins suggested, childhood was seen as a crucial time for psychological development. Most psychologists presented emotional growth as orderly, predictable, sequential. They made it sound simple. Presumably that was part of self-esteem's rising appeal. They also forgot about the marginalized. Perhaps, for some readers, this was part of its appeal too. In the context of a conservative backlash, self-esteem literature written for parents placed new onuses and pressures on mothers to optimize their children's emotional wellbeing. However, as this chapter argues, African American readers viewed white self-help strategies as oblivious to their particular needs and, as an alternative, continued to develop their own projects and initiatives. For them, building self-respect and self-esteem among children was a matter of urgency, part of their survival even, not a middle-class luxury, as Black children needed enough confidence to hold their own in a still racist society.

Emotional growth and the American family

In the twentieth century, childhood was increasingly treated as a major social, medical, and political issue. Historians refer to the "century of the child." Developments in doctors' ability to cure many infectious diseases caused a sharp decline in childhood mortality and illness, due largely to the arrival of antibiotics and other curative drugs. But despite improvements in the physical treatment of children, concern about their general wellbeing endured. Arguably, it intensified. In contrast to the rough and ready Victorian times, twentieth-century society saw children as emotionally vulnerable, fragile, readily overburdened, and in need of careful handling. Waning concern about physical health gave way to growing apprehension about emotional wellbeing. If a majority of children, for the first time in human history, were likely to survive into and enjoy a long adulthood, then it seemed crucial to grant them an appropriate emotional start in life.[12] Whereas Victorian children were "priceless," in the twentieth century, declining infant mortality rates and smaller families made each child more "precious."[13]

The immediate postwar period saw considerable change in American family life. American youth rushed into marriage and childbearing typically in their late teens or early twenties. In the early Cold War period, family stability provided a comforting bulwark against the threats of communism and nuclear war. The white, middle-class suburban nuclear family was a product of the 1950s, one guided by a new breed of medical, psychological, and scientific experts, emerging around the very same time that families and individuals became more isolated. Modern society urged families to jettison their wider familial networks to become "nuclear" units specializing in emotional nurturance, childrearing, and raising democratically minded citizens.[14] As one self-esteem author, Jean Illsley Clarke, suggested in 1978: "When we give up extended families, we gave up one kind of support system and parents in nuclear families are not assured of a support system."[15] In these new support systems, experts replaced the traditional role of relatives.

At the same time, feminists contested the nuclear family and ideas such as maternal instinct, viewing these suspiciously as social fabrications that served patriarchal needs. They deconstructed the modern family to expose it not as natural, but as an artificially constructed institution that sustained patriarchal forces. French feminist Elizabeth Badinter denied altogether that maternal love was biologically determined.[16] Counterculture activists also warmed to antipsychiatry ideas which insisted that mental illnesses, especially schizophrenia, had familial causes, as this further helped legitimate their revolt against the nuclear family.[17]

Phyllis Schlafly was far from amused by this feminist approach. From the mid-1970s, the neoconservative backlash strove to reinstate the primacy of the nuclear family as the basis of American morality and social order. Rising divorce levels made this seem all the more urgent.[18] Mother-blaming had a long history, but became a staple part of conservative ideology in the 1970s and 1980s. According to historian Ruth Feldstein, "maternal failure, social and emotional pathology, and damaged citizens became the mantra for anti-welfare, anti-civil rights, and anti-feminist postures." These perspectives permeated mainstream self-esteem advice, which advocated working on emotional wellbeing at an individual or familial level, and not much farther beyond.[19]

Sociologist Frank Furedi writes:

> The contemporary therapeutic imagination is haunted by the belief that damage to the emotions is systematically inflicted on the individual within the family and during the course of day-to-day interpersonal relations. Consequently, therapeutic culture conveys a strong sense of unease toward the private sphere. At times, this unease turns into hostility toward the informal world of private life.[20]

Similarly, historian Deborah Weinstein argues that late twentieth-century concern about "pathological families" was a central aspect of therapeutic culture's expansion. Therapists approached families as a single organism in which one "sick" member might "infect" the entire unit with their emotional pathologies.[21] For that reason, homes needed to offer healthy emotional environments.

Emotions were crucial to this new vision of stable family life. No historical precedent existed for the burgeoning emphasis on producing a world of satisfaction, amusement, and inventiveness for children in small, contained nuclear units. Never before had parents (usually mothers) been under such pressure to energize children with an emotionally expressive personal life. True, nineteenth-century middle-class families had been urged to treat the domestic sphere as a haven and retreat from the pressures of industrial, modern life, a private area of domestic tranquility in which children were (ideally) showered with positive emotions such as love. But attitudes had changed by the late twentieth century. Whereas identities had once been forged by occupation, postwar identities were defined more strongly by personality. In an era increasingly dominated by capitalism, this fed well into conservative values that sanctified the home as the place where democratic, individualistic, and competitive citizens were formed.[22] Of course, domestic ideals often differed from lived realities. Indeed, the construction of the ideal nuclear family most likely deepened anxieties among those who failed to live up to the new expectations. Nonetheless, the ideal was powerful, persuasive, and influential.

Social psychiatrists obsessed over the interpersonal dynamics of family life to understand and manage emotionally unhealthy families.[23] In the 1970s, many psychologists believed that too many Americans had been inadequately nurtured in their youth, the consequence of which was a rising tide of social problems caused by people with defective personalities. Some social psychologists warned that sufferers of low self-esteem risked developing authoritarian personalities better suited to countries beyond the Iron Curtain. In historian Michael Staub's words: "If the psyche of an individual was vulnerable while young, and if there existed a decisive link between the development of personality and the evolution of ideological values, then environments (particularly within families) required close scrutiny."[24]

In this sociopolitical context, family environments were subject to expert inspection. Social psychologists positioned themselves almost as rescue services, extricating children from destructive family relationships.[25] They approached childhood as a phase of life in which harmful psychological habits were too easily ingrained, and when emotional wellbeing was threatened by numerous forces: divorce, inattentive parents, school bullies, domineering teachers, the turmoil of adolescence. They warned of the potential consequences for adulthood: a possible future of delinquency, crime, and sexual immorality. But hardly ever did they reflect on the influence of race, gender, sexuality, and class in shaping emotions, psychologies, and personalities, for these subjects were now seemingly forbidden.

Life's emotional foundations

In a 1972 *New York Times* article on Harry Stack Sullivan acknowledging his newfound popularity some decades after his death, parents were advised that simply attending to their infants' physical needs (e.g. food) was not enough. To aid their healthy affective development, infants needed to be mothered, held, rocked, talked to, and touched. If emotional care was lacking, infants might withdraw, become apathetic, and "even literally shrivel up and die." The damage done could be irreparable.[26] Similarly, in 1969, Ross Laboratories, a baby-food company, produced a pamphlet: *Developing Self-Esteem*. Although primarily concerned with promoting baby food and prenatal vitamins, the pamphlet carefully considered the importance of emotional growth. Just as a baby's physical growth depended upon food, emotional development relied upon positive relationships and interpersonal interactions. As the pamphlet explained: "The ways in which we were regarded in those early years determines to a great extent the esteem we now have of ourselves as adults. The feeling that we have dignity and worthwhileness as persons provides a healthy soil for the growth of self-confidence." However, the flipside of this coin was that belittling attitudes could "plague us unhappily throughout life."[27]

Although earlier psychoanalysts, influenced by Freud, had stressed the lifelong consequences of negative infant experiences, postwar psychologists contemplated emotional growth more comprehensively, also drawing attention to the positives. In the 1950s, Erik Erikson's influential psychosocial model suggested that infanthood was the key time to mold personalities, and that positive interactions between parent and child were crucial to the making of America's future citizens.

Erikson advised parents against accidentally cultivating psychological distress. Instead, children needed to be shielded from behaviors that encouraged "abnormal" emotional growth.[28] Evidently, families now had new pressures placed upon them to deliver supportive emotional foundations.

In turn, experts adopted new roles. They no longer presumed that women somehow knew instinctively how to raise their children, or that they had regular support from extended family members. In they stepped to offer guidance.[29] British psychologist John Bowlby warned that children raised without a strong maternal bond would likely grow up delinquent. Bowlby's influential "attachment theory" stressed the importance of emotional bonding between mother and child, typically between six and twenty-four months, as the foundation of a healthy emotional adulthood. Although Bowlby's work met with mixed responses in the psychoanalytic community, his ideas garnered considerable public interest. His impact on childrearing in America was remarkable; he helped cement the view that positive emotional development depended largely on maternal love.[30] Enormous ideological power was placed on maternal influence. Despite feminist remonstrations, motherhood was still widely considered as *the* defining aspect of a woman's identity. As a general rule, fathers weren't expected to bear much responsibility for a child's emotional growth. Women, after all, were stereotypically associated with emotion, thereby placing responsibility firmly onto them. In this context, many mothers turned eagerly to the popular psychology and childrearing guides being published by experts.

In turn, a burgeoning literary genre emerged that focused on self-esteem and parenting. The new style of books replaced strict discipline with emotional nurturance and warmth, and promoted fun forms of childrearing; this was in reaction to wartime austerity, military discipline, and earlier eras of stern childraising. Dour topics, such as eugenics, were avoided and parents were instead urged to enjoy being with their children.[31] Parents were encouraged to create positive emotional spaces. Stanley Coopersmith warned that "children reared under such crippling conditions [excessive discipline] are unlikely to be realistic and effective in their everyday functioning and are more likely to manifest deviant behavior patterns." Coopersmith urged parents to accept their children, establish clearly defined behavioral limits, and respect individual initiative.

This, he hoped, would create positive self-evaluations and increase self-esteem.[32] Of course, one should not presume that the ubiquity of baby books meant that mothers accepted their content entirely without question. Nonetheless, a particular ideological stance was undoubtedly articulated and established within their pages.[33]

Psychological intervention was directed at healthy and sick children alike, another manifestation of psychology's shift away from exploring psychopathologies alone. One author, Shirley C. Samuels, stated in the introduction to her 1977 guide to self-concept: "This book is intended to help the parent and teacher with the child who is not severely emotionally disturbed, and to recognize that further help is needed for the one who is."[34] Child experts now considered it within their purview to guide the emotional growth of all children, influenced by psychologists including Anna Freud who examined both childhood pathology and normality.[35]

However, this subgenre of self-esteem literature essentially spoke to the idealized, middle-class, nuclear family in which mother, opting not to work, had enough time, patience, and energy to devote to her children. This was not a genre that busier mothers, forced to work through economic necessity, would have readily identified with. Marginalized families, still shunned from education and workplaces, formed a major part of that less affluent socioeconomic group. According to the experts, parental indifference was out of the question, but the middle-class psychologists who were insisting on such behavior were largely sheltered from the harsh realities of American family life. Morris Rosenberg condemned parental apathy (or, from another perspective, limited resources for time-consuming, emotionally intensive baby care) for harming children's self-esteem. He considered disinterest to be more dangerous than excessive punishment.[36]

Benjamin Spock was the expert most associated with permissive attentive childraising. For decades, his book, *The Common Sense Book of Baby and Child Care*, first published in 1946, was a mainstay of middle-class homes. Spock championed the idea that childrearing should be fun, albeit emotionally intensive. Recalling his own childhood, he remembered his mother as tyrannical, moralistic, and opinionated, although extremely dedicated to her children. When writing his own childrearing book, he targeted it at war-weary parents who, like himself, resented the harsh, discipline-heavy parenting they themselves had experienced. As a welcome alternative, Spock suggested that parents pick up their babies, give them affection, and enjoy their company. His was a psychoanalytic approach teaching parents that their actions early on in a child's life had potentially enormous consequences.

Recalling his own experiences of low self-esteem, Spock wrote:

> I find it easier to emphasize and clarify aspects of low self-esteem and its causes, perhaps because my mother was particularly eager to ward off conceit in her children, which she thought of as obnoxious in itself and likely to lead to more serious faults. I still recall today how, when one of her friends complimented me on my looks in adolescence, she hastened to say, after the friend left, "Benny, you are not good looking. You just have a pleasant smile." All six of her children grew up feeling somewhat unattractive and under-achieving …
>
> … As a child I was so often reprimanded for "naughty" acts that every time I come home from playing or from school, I felt a cloud of guilt around my shoulders. "What have I done wrong?" I would wonder, when actually I had rarely done anything wrong. I was a goody-goody. This was because my mother was so critical, so on the lookout for misdemeanors, that my five siblings and I felt chronically guilty … I am describing how it worked in our family only to make the point that in the matter of self-esteem the first and most important step, I feel, is not so much to build it up with a succession of compliments as to avoid tearing down the natural self-assurance with which children are born. Then you don't have to run the risks of conceit and excessive self-satisfaction that can result from too much praise.[37]

The emerging self-esteem advice argued that infants develop their social and cognitive selves in their formative years. In *Your Child's Self-Esteem: The Key to His Life* (1970), Dorothy Corkville Briggs informed readers that babies were not born with a sense of self. Nor did they instinctively develop a sense of selfhood. Instead, positive self-esteem was learnt from living and interacting with others.[38] The influence of parent–child interactions was felt immediately from birth, imbuing parents with duties to inculcate self-esteem from day one. "During the first year, consistent loving care leads to a sense of trust which is the foundation of identity," added Shirley C. Samuels; "the healthy relationship with the caretaker enables children to begin the process of separating themselves from the caretaker ... positive self-concept in all its dimensions will result if trust, autonomy and initiative are appropriately encouraged."[39]

The new parenting style required time, patience, and care. While waiting for infants to learn to speak, nonverbal cues were important: caressing, singing, cuddling, touching, rocking. Touching a baby's skin was thought to stimulate both physical and psychological growth.[40] Simple, positive interactions such as these had long-term benefits. As John Mack and Steven Ablon argued in their 1983 book on developing and sustaining childhood self-esteem:

> One might consider the precursor of positive self-esteem to be the good body feeling and the effects of pleasurable excitement and lively interest associated with the mutually loving exchanges, the touching and looking, that occur between the parent and the infant. It is for this reason, one may suppose, that self-esteem remains so intimately linked throughout human life with our valuations of the attractiveness of our bodies and faces and with physical wellbeing more generally.[41]

Without these physical expressions of love, babies might experience depression, appetite loss, weight loss, and even death by marasmus (or wasting away), so parents were warned. Ominously, Clarke advised that a lack of touching early in life eventually meant that, later in life, there is more chance that someone might "end her life quickly with an automobile or slowly with alcohol or other drugs or [s]he may live her life without really being, by living only a small part of the life [s]he is capable of living. There's little positive self-esteem there."[42]

Bad parents?

It is worth pausing here to reflect upon Clarke's comments, for they render visible the double-edged nature of self-esteem advice. On the surface, self-help was positive and helpful. In reality, it was underpinned with portentous, morally persuasive, encoded messages. Rarely did it shy away from alerting readers to the dire implications of failing to heed the expert's advice. At worst, self-esteem books were threatening, moralistic, and critical of those unwilling – or simply unable – to emotionally nurture their child. They encouraged mothers to doubt their parenting skills and ability to fully support their children without being watched over by experts. They enlisted mothers to center their lives around infants, while constantly self-monitoring their own behavior.[43]

At worst, the self-esteem movement began nurturing a dependence on self-help books that fed off parental insecurities and personal traumas. Self-esteem guidance started to victimize "inadequate" students, "bad" parents, and "mean" teachers, and then offered therapeutic solutions to resolve those apparent flaws.[44] As a business, it was hugely successful in helping create the very problems it set out to resolve. Certainly, Clarke left readers in little doubt about the calamitous outcomes of failing to infuse children with self-esteem. Positive and supportive on the surface, self-esteem advice was awash with moral messages about parents' personal and social duties.

Here was capitalist parenting at work, urging mother to discipline herself, remedy her allegedly deficient parenting techniques, and shoulder the blame should all go wrong. Here, again we also see the workings of "bio-morality": the insistence that working on oneself is both a moral duty and a social obligation. Earlier in the twentieth century, the Child Guidance Movement, a psychiatric initiative, had been notoriously critical of mothers. Underneath its benevolent guise, it blamed negligent mothers for each and every fault perceptible in their children, and for epidemics of bad youthful behavior.[45] The fashion for mother-blaming translated easily into postwar publications on parenting, infants, and self-esteem.

During the 1970s, Briggs wrote a number of popular books on self-esteem, including the aforementioned *Your Child's Self-Esteem: The Key to Life*. Like many of her contemporaries, Briggs castigated earlier psychologists for being infatuated with psychopathological disorder. As an alternative, she paid attention to children's psychological needs, healthy or otherwise, which enabled her to expand her remit away from the clinic to more intimate spaces of American familial life. Briggs commenced her book by warning that too many people "limp through life in inner turmoil" and that "neurotic hang-ups have become a way of life." She also commented on the paradox of children being America's most important national resource even as parents had limited knowledge of how to successfully raise them.[46]

Briggs also argued, with much moral persuasion, that her self-esteem-raising methods could help children avoid a future of delinquency, alcoholism, and nervous breakdowns.[47] Hardly reassuringly, she reminded parents:

> In spite of good intentions and heartfelt efforts, however, some of you find your youngster not turning out as you would like. He is underachieving, emotionally immature, rebellious, or unduly withdrawn ...
>
> Even if your children are not having problems, hearing about the rising rates of juvenile delinquency, dope addiction, drop-outs, venereal disease and illegitimacy hardly lowers your anxiety level ...
>
> If you are like most parents, your hopes for your children are based on more than their avoiding nervous breakdowns, alcoholism, or delinquency. You want life's positives for them: inner confidence, a sense of purpose and involvement, meaningful, constructive relationships with others, success at school and in work. Most of all – happiness.[48]

Briggs's message was clear. Inculcating self-esteem in children might well be optional, but failure to do so could impact catastrophically on both child and society. Inculcating self-esteem was therefore a sociomoral obligation. As she added: "Self-esteem is the mainspring that slates every child for success or failure as a human being. The importance of self-esteem in your child's life can hardly be over-emphasized. As a parent who cares, you must help your youngsters to a firm and wholehearted belief in themselves."[49] Briggs defined self-esteem as how much a child liked themself. It should not be "noisy conceit" ("a whitewash to cover low self-esteem"), but a "quiet sense of self-respect, a feeling of self-worth." "When you have it deep inside," explained Briggs, "you're glad you're you."[50] Self-esteem made children "fully functioning." For Briggs, the self-judgments made by children in their formative years later influenced their choice of friends, marital partner, whether they would lead or follow, and so on. For that reason, self-worth needed to be ingrained into personalities at the start of life. Like many social psychologists, Briggs looked toward interpersonal dynamics, not social structures, collective action, or government interventions. Children, presented in self-help literature as a homogenous group, were born with potential for full emotional growth, regardless of material or social circumstances or social injustices facing them. Life's success depended upon the emotional climate of their homes.[51]

Briggs was not alone in linking personal self-esteem to the welfare of American society. Mack and Ablon presented self-esteem as a crucial personality determinant. Self-esteem guided everyone while also impacting collectively upon social functioning. The authors warned portentously that "the child who does not experience his value in the eyes of his parents will fear their abandonment and its attendant dangers to his existence. The older child and adolescent may even contemplate suicide."[52] Mack and Ablon conceded that some adults could mend damage to self-esteem inflicted in early years, but this could never be a complete repair, especially if the early hurt had been severe or prolonged. Any repairing carried out in adulthood would be partial, and childhood wounds could be easily reopened in the process.[53]

In light of such statements, it becomes apparent why critics have accused self-help advice (and psychologists) of victimizing mothers and feeding off parental insecurities. In the 1970s and 1980s, faith in parents was low. So low, in fact, that failing parents were themselves accused of lacking self-esteem. Parents, too, were now expected to reach psychological maturity before bringing harmony to the nuclear family. Inability to sustain a happy marriage or raise a "normal" child were portentous signs of arrested emotional development. Self-esteem advice encompassed the entire nuclear family, a unit requiring effective emotional nurturance to function effectively.[54]

Parent–child relationships were understood as open, interactive systems, replete with reciprocal interpersonal interactions within the unit. If children suffered low self-esteem, this might rub off on the parents, establishing an undesirable Horney-esque vicious cycle. How children responded to parental attention in turn impacted on the parents' self-image, their feelings of competence, their confidence in themselves as parents.[55] Self-esteem flourished only with mutually beneficial reciprocity. As one author explained: "The reciprocity between husband and wife bears directly on the parenting experience. The husband supports and nurtures his wife so that she can care for the infant without becoming depleted."[56]

In the late 1970s and 1980s, Clarke's book was read eagerly by parents and families across America, Canada, and Europe. Clarke claimed to have written it in response to her own feelings of parental inadequacy. As she explained: "I needed better ways to take care of my children and myself, to build positive self-esteem in my family. I began to meet with groups of people who also wanted better ways to build self-esteem in their families." The assumption here was that it was natural, even common, to feel underconfident about parental ability, but, not to worry, experts were now here to help. Clarke argued that parents had a "responsibility" to improve their self-esteem. "Love yourself, appreciate your accomplishments," urged Clarke. "Accept your mistakes and change them without beating on yourself. Find some other people who will swap that kind of healthy relationship with you."[57]

The "other" families

In adopting a one-size-fits-all approach, self-esteem advice aimed at parents proved itself oblivious to the sheer diversity, and varying needs, of American families. Even white working-class readers often failed to recognize themselves or their circumstances in this literature. Black readers felt entirely excluded. The great irony was that Black working-class families were considered a major social problem. Thinking logically, one might have expected self-esteem books to target those families deemed the most troublesome. But they didn't. Rather, they proved more intent on sustaining idealized images of white nuclear families than on realistically tackling social issues. They were color-blind and indifferent.

In 1965, Daniel Patrick Moynihan published his notorious *The Moynihan Report: The Case for National Action*. At the time of publication, Moynihan was a 38-year-old liberal politician and academic. He wanted the administration to understand how powerful forces, especially unemployment and poverty, devastated lower-class Black families. An Irish Catholic from New York, Moynihan considered himself especially mindful of the enduring power of religious, ethnic, and racial identifications.[58] Moynihan's report was released in the same year that Kenneth B. Clark published *Dark Ghetto*, another psychologically informed book on race. Clark avoided portraying African Americans as infantile or without hope, but did insist that socioeconomic distress caused grievous psychological harm and a culture of deprivation. Indeed, he coined the phrase "tangle of pathology," which Moynihan then controversially used.⁵⁹ Moynihan's report warned that the "negro family" was crumbling. Middle-class Black families had saved themselves, but left behind those further down the social scale. Moynihan wrote that "at the heart of the deterioration of the fabric of negro society is the deterioration of the negro family." The family was the fundamental source of weakness in Black communities. In dysfunctional Black families, centuries of historical injustices were still manifest. Unless family sanctity was restored, efforts to end discrimination, poverty, and injustice would fail, for the family was America's basic socializing unit. Traumatic childhood experiences might cause lasting psychological damage.⁶⁰

As evidence, Moynihan warned that nearly a quarter of urban negro marriages were being dissolved; nearly a quarter of negro births were illegitimate; almost a quarter of negro families were headed by single mothers; and the breakdown of the negro family had caused widespread welfare dependency.⁶¹ Of particular note were Moynihan's views on Black women being forced to go out to work because their husbands were not earning enough money. Black families had been forced into a matriarchal structure, the converse of the white nuclear family in which the mother ideally stayed at home. The majority group was operating on one principle; the minority group on another. Black progress was held back by the crushing burdens this imposed on the male.⁶² Although vague on solutions, Moynihan wanted the state to develop initiatives that would strengthen Black families, enhance their stability and resources, and "bring the negro American to full and equal sharing in the responsibilities and rewards of citizens."⁶³

The Moynihan Report became one of the most controversial social research pieces ever published. Critics raised various points. Why precisely was a matriarchal family at a disadvantage? Why had Moynihan pushed discrimination and racism to the background? Indeed, Moynihan seemed to be suggesting that, even if racism ended, Black problems would still not be solved.⁶⁴ Some accused Moynihan of oversimplifying, overdramatizing, and downplaying the diversity of Black communities. Patricia Harris, a Black leader, later observed that family issues became *the* problem, not just one of many problems, despite the ongoing presence of white discrimination.⁶⁵

Shortly before the report was released publicly, the Watts Riots broke out in Los Angeles, sparked by the arrest of Marquette Frye, a 21-year-old

African American man which led to six days of civil arrest and allegations of police abuse. Thirty-four people died, and 14,000 members of the California Army National Guard suppressed the disturbance. Many civil rights leaders backed away from the report, partly because of the riots. While recognizing some truths, they regarded an agenda based on pathologies and stereotypical Black welfare mothers as unhelpful. The true pathology, they retorted, was racism; familial problems were simply racism's scars. Moreover, none of the leaders knew Moynihan too well. Nor had he consulted with them. Even those who respected Moynihan's intellect and desire for reform shrank away. Martin Luther King feared that emphasizing the negatives of Black life would ultimately strengthen accusations of innate negro weakness.[66] Moynihan was a liberal; he viewed his ideas as progressive and advocated equal opportunity and government action. However, in reality, liberal and conservative thought converged in the report by refusing to see Black problems as systemic in American society.[67]

A stream of sociological studies set out to prove Moynihan wrong and demonstrate that Black families weren't so very different from white ones after all.[68] In 1972, Robert Hill, director of National Urban League, published a report that emphasized Black strengths, using the same data sources as Moynihan. This report found little evidence that Black families were pathological.[69] No longer did Black families tend to be presented as a social problem, rife with illegitimacy, broken homes, and men suffering low self-esteem due to their matriarchal home life. Such descriptions became reassessed as myths, and Blacks were depicted as quite like whites after all.[70] This positive turn produced a curious research finding: self-esteem levels among Black and white children were not as different as had often been presumed. Something seemed to exist in Black culture, unnoticed in earlier models, that shielded the self-image of Black youth from white racism's constant onslaught. In his more flattering 1968 book, *Black Families in White America*, Andrew Billingsley wrote that being in a "broken family" usually impacted negatively on white children's self-esteem but, for reasons unknown, this was less of a problem for Black children. Perhaps Black culture didn't stigmatize broken families as harshly as white culture did.[71]

Other research produced even more mysteries. In 1972, after finding little difference in levels of Black and white self-esteem, Morris Rosenberg and Roberta G. Simmons speculated that Black families were distinct in that it was only when entering a white environment and being made to feel different, abnormal, or ashamed of the "Black broken home" background that self-esteem began to crumble.[72] Well into the 1980s, African American social scientists insisted that children's experiences varied immensely, but, nonetheless, Black children's unique socioemotional environment did not necessarily seem consistently to lessen their self-worth or create an overwhelming sense of self-hatred. Positive self-worth was somehow maintained.[73] Of course, as discussed in the previous chapter, such findings may have arisen due to a fear among researchers in the early 1970s of publishing negative findings that could be interpreted as racist or discriminatory. Positive findings on race were politically safer.

Some argued that Black families were emotionally stronger than white ones. In 1971, Joyce A. Ladner argued that Black children enjoyed greater emotional stability than white children, and enjoyed better integrated personalities that their white, middle-class counterparts. The emotional mollycoddling going on in middle-classes families produced fragile personalities and protected, sheltered lives.[74] Similar sentiments were expressed by Lee Rainwater:

> Lower-class negro women do not show the deep psychological involvement with infants and young children that is characteristic of higher social classes. They rarely manifest the anxious attention to children, the sense of awesome responsibility along with the pleasure, that is characteristic of many working-class women ... Taking care of babies is regarded as a routine activity which is not at all problematic. There is no great concern with the child's development, little anxiety about the appearance, crawling, talking, walking and other signs of growth and maturation. These things come as they come, with little more than passing comment on the part of the mothers and others who care for the children.[75]

Critics also questioned the assumed cultural homogeneity between Blacks and whites. In their 1975 study of Harlem, Barry Silverstein and Ronald Krate argued that the child-centered orientation of middle-class white America, with its reverence of a long, protected childhood, developed in the postwar age of affluence. Yet the poverty in which many urban Blacks lived hardly encouraged such sentimentality. For working-class Black communities, life had harsh realities. Children needed to be socialized to survive hardship and endure racism. Permissive, child-centered approaches involving stay-at-home mothers catering constantly to the emotional needs of her infants were simply not realistic. Persistently high infant mortality rates, disease, and hunger among Black children suggested to some that many Black families still struggled to attend to their infants' basic, let alone nonbasic, needs.[76]

Throughout the 1970s and 1980s, research into child development pursued by African American theorists usually concluded that both self-help and white social science had little relevance for Black families. Minority scholars now studied minority children from minority culture perspectives, recognizing that white research and concepts contained hidden Eurocentric values: individualism, competitiveness, futurism, and objectivity. In contrast, Black culture, broadly speaking, was presented as shaped by spirituality, harmony, expressive individualism, orality. Parents blended together white and Afrocentric socialization. The idea of biculturality gained appeal among African American communities: the idea that African American children, by necessity, were forced to develop the skills required to survive and succeed in both Black and white America.[77]

Gloria J. Powell and Marielle Fuller's 1973 study, *Black Monday's Children*, explored the effects of school desegregation on Southern children's self-concept. It recognized that Black families were far more likely to be on the lowest socioeconomic rung and that, consequently, Black families were likely to be less stable. In the authors' words: "Thus the emotional and attitudinal climate into which a Black child is born may be one of a deeply ingrained impoverishment."[78] Psychologically, Black children might have unmet dependency needs and fail to receive the proper basis for accruing self-esteem while young. At school, racist attitudes of white teachers toward Black children worsened the problem further. As Powell and Fuller pleaded: "Somehow, somewhere, sometime the American people must wake up to the great American nightmare and the sanguine unreality of the American dream. We are a racist society, and until we have solved that very basic problem school desegregation may produce more problems than it will solve." As African American parents, Powell and Fuller suggested that, since public school education was failing Black children, it might be best for Black communities to educate their own, bypassing debates on bussing and segregation.[79]

Aside from cultural differences, racism was a key structural problem identified usually only in Black-penned self-help literature. Alvin Poussaint was a key spokesperson for the idea that Black children were being socialized in damaging environments of prejudice and discrimination. Black and white parents shared an imperative to provide for their children, both physically and emotionally, but Black families, particularly mothers, had the additional dilemma of raising their children in a racist society (known as "racial socialization"). In 1972, Poussaint asked:

> Is it possible to raise a psychologically unscathed Black child in a racist environment? How do we go about it? What can we as Black parents do to help our children grow into strong and dignified men and women? In the struggle with problems like these, the abundance of child rearing information directed to the white person is of little use to Black families.[80]

Poussaint perceptively recognized that parental self-help advice had only white, suburban, nuclear families in mind, despite society seeing Black families as the problem. Like other authors of self-esteem books, he certainly saw great value in raising children in emotionally supportive environments, but attention needed to be paid to culturally specific needs. Black parents had to ensure that racial dignity was more than rhetoric. Leading by example, they could help build a strong sense of community. The fruits of Black power could be ingrained into the next generation of African Americans, who would know their own history, appreciate their heritage, and fight for their rights.[81]

In practice, Poussaint advised Black parents against making derogatory comments about their own culture, in case this impacted negatively on their child's self-esteem. Self-prejudice was too easily absorbed: "To base a child's self-esteem on a lack of respect for others is to make him vulnerable to the insidiously racist thinking that already pervades this society. The Black child will be more likely to develop a sound sense of his own worth in a family where the dignity of all people is respected."[82] Of course, it was necessary for white parents to raise their children not to be racist, and Poussaint did indeed devote a chapter to this subject. Yet, this was perhaps unlikely to happen, and such matters never figured in mainstream self-help advice. Poussaint also considered it necessary for Black children to hold positive attitudes toward each other. Parenting styles needed to be attuned to ethnic pride; children had to be encouraged to feel comfortable in their colored skins. This would buttress against racism's emotional harm.[83] Evidently, African American conceptions of self-esteem were far more attuned to structural realties than their (more popular) white self-help equivalents.

In 1975, Poussaint, along with child psychiatrist James P. Comer, published a wide-ranging study in which they criticized parental advice books for being written first and foremost for white middle-class families. In more than 400 pages, written in a question-and-answer format, they explained the basics of childcare to Black parents, while also addressing the uniqueness of their situation. It seemed to the authors that most child psychologists presumed that Black and white children were more or less the same.[84]

These African American authors all recognized that Black children had specific developmental needs. Some were cultural, but the majority were caused by structural issues and social discrimination. Largely excluded from mainstream self-help research, Black theorists developed their own reflections on racism's impact on child development and ways of salvaging the socioemotional and psychological growth of Black children. Comer and Poussaint believed that African Americans needed to preserve their culture and communities, but worried about whether "white psychology and childrearing approaches will change us, hurt us, destroy our culture." While they acknowledged the limitations of most parenting guides, even for white working-class people, they identified racial matters as acutely problematic for working-class Black families. Comer and Poussaint undoubtedly shared with white authors a belief that children, from the day of birth, needed love, care, and training. However, the aim was slightly different: "With such preparation, Black children are able to maintain positive feelings about themselves when growing up in racist environments, whereas those without proper preparation often have low levels of self-esteem." The authors advocated building a positive Black identity based on "an inner core of pride and positive feelings, or it may fade away under the harsh light of life's realities."[85]

Comer and Poussaint knew full well that Black children were likely to experience and express negative emotions, and sought to help them channel these in constructive ways. To provide an example, they were fairly tolerant of anger, but only if channeled into "socially acceptable, useful energy." Black children needed to learn inner control and act aggressively under duress. Being super-sensitive to injustice was an inadequate solution if it encouraged children to be indiscriminately reactive and angry.[86]

Other books written for Black parents included Phyllis Harrison-Ross and Barbara Wyden's *The Black Child* (1973), which, across its 350 pages, adopted a case-study, storytelling approach. Like other contemporary child guidance books, it discouraged harsh discipline and punishment and underscored the importance of self-esteem to emotional wellbeing. Harrison-Ross and Wyden were particularly outspoken about the social, economic, emotional, and educational deprivation endured by Black working-class families. They feared that many children were being emotionally trained in wrong ways, for "so many Black parents have lived with hate and fear that they carry invisible, unsuspecting forces in themselves that perpetuate that heritage of hate and fear in their children." Parents needed to become aware of these "emotional booby traps" and reverse a vicious circle. In the author's own words:

> We live in a racist world where almost every minority has to fight to establish and maintain its self-esteem. The Black minority has to try harder than most, because its self-image has been cruelly distorted by the harsh inequities of history and of our present social and economic environment ... parents need to recognize their own psychic scars of being treated as second-class beings.[87]

A core aim of this book was to help Black parents counteract the pressures of a racist society on the upbringing of their children, and to help rear their children with a proud sense of identity.[88]

In 1978, African American psychologist Amos N. Wilson aimed similar criticisms at psychologists who treated Black and white children alike. In his view, psychologists believed that "the Black child is a white child who happens to be painted Black," with white middle-class children being upheld as a superior model that Black children ought to emulate. The developmental patterns of white middle-class children were seen as the optimal standard. Yet, as Wilson concluded, "to ignore race and its many ramifications when dealing with the Black child's educational problems, emotional problems, or whatever is tantamount to ignoring the child himself and yet this is exactly what American psychology does."[89]

While it is often hard to find the voices of parents themselves in the historical record, some insight can be gleaned from Marie Ferguson Peters's 1985 Toddler Infant Experiences Survey (TIES), which foregrounded self-esteem as an important indicator of the nature of child development. Peters found that self-esteem was still of prime importance to Black families and was even, unlike in middle-class white families, considered essential for sheer survival. Black parents knew that the day would come when their unsuspecting toddler would grow up and experience racism for the first time. Childrearing was heavily informed by this sense of impending racial doom. The sixteen parents surveyed by Peters all mentioned the importance of their children learning to "be positive about themselves," of daughters having "respect for herself" and sons getting "respect for being himself." One mother commented: "I'd like them to have enough pride, because if you have enough pride or self-confidence in yourself, you'll let a lot of things roll off your back." Another parent stated that she wanted her children "to feel sure about themselves. They need this to be able to make the best of their future." Others said: "I want them to be proud of the fact that they are Black"; "I tell my child, 'You're Black, we're all Black in this family, be proud of it!'"; "I've told my daughter many times, 'You're a pretty little Black girl and I'm proud of you!'"; "I tell them to hold their heads up and not to be ashamed to be Black."[90] (Sadly, Black female authors such as Mikki Kendall, writing in 2021, still discuss the persistence of "survival parenting" in a hostile, white America).[91]

The general consensus in Black-penned literature was that self-esteem was a crucial buttress against white-dominated society. Black children's socialization took place in a very specific environment, one informed by both Black and white culture, as well as the more general force of capitalism.[92] Black professionals had a particularly important role to play from within. Robert C. Johnson was one Black psychologist who advocated childrearing techniques that fostered positive pro-Black attitudes and behaviors, and introduced these to parents in ways that were systematic, enduring, effective, and compatible with Black American life. Johnson thought that mothers should instill in future Black citizens a sense of self-respect and positive views about other Black people. Socializing infants and children to absorb and disseminate pro-Black values would strengthen Black communities and, in turn, increase opportunities for securing economic resources. But what were these values? For Johnson, these were the ancient values of Africa, the ones that guided African forefathers through slavery and segregation. They were the values of communalism: respect of the group and self, self-knowledge, group perpetuation, spirituality, humaneness, and humanitarian concern.[93]

Anderson J. Franklin and Nancy Boyd-Franklin, discussed at the start of this chapter, similarly believed that Black parents ought to be made more aware of the psycho-educational process and participate in it in a meaningful way, fully acknowledging the additional pressures that face Black families living in a racist society. "It is only through this process that Black children can have a chance for a viable future and an opportunity to realize their potential as adults," they wrote.[94] The construction of alternative perspectives on self-esteem was symbolically important, for it recognized the fundamental whiteness of mainstream self-esteem advice and its construction of idealized families that were essentially white and middle-class. Moreover, Black communities presumably benefited from the consciousness-raising of the late 1960s that had strikingly emphasized the importance of self-esteem and the need to embed it in all aspects of Black American life. Perhaps they benefited too from avoiding the excesses of intensive parenting methods, later encapsulated by the term "helicopter parenting," in which parents become overly focused on their children in ways that might not necessarily be healthy or beneficial for either parent or child. The term stems from the idea that helicopters always hover overhead, constantly overseeing every aspects of their child's life. It was already in use by the 1980s, suggesting some skepticism early on to the amount of attention that white, middle-class America was urged to pay to childrearing.

It should seem clear to readers by now that mainstream self-esteem advice mostly refrained from talking directly to so-called "problem families," and instead adopted an approach that upheld certain families (typically white, middle-class ones), as an ideal that all should follow. Central to this was the idea of constructing particular personality types that reflected these cultural preferences. Yet the personality type to be forged was not usually Black, gay, a hippy, a rebel, or anything else deemed undesirable in postwar American society. The existence of social groups beyond the American "norm" was not attacked or maligned, but simply ignored. The chapters that follow continue to explore these themes by progressing onward throughout the life cycle to examine school children and then teenagers.

FIVE

School

The first day of kindergarten is a milestone in the lives of children, a day when they begin to forge ahead toward independence in a new environment. Parents and family will no longer be the only primary persons in the child's life. Others – those teachers, friends, and peers whom a child considers significant – begin to influence and affect development of the child's self-concept. If these new significant others consider and treat the child as a worthwhile and an important human being, they will help the child develop a positive self-image.
Michele and Craig Borba, *Self-Esteem: A Classroom Affair*[1]

This was the first time I really knew that I was different and that somebody would be mean to me because of it. Consider the impressionable minds of five-year-old children and the realization that skin color made Karen and me the subject of disdain … On that day, I changed from a carefree little girl to a cautious and insecure one, not being sure when somebody might be mean to me again because of the color of my skin. The realization that I might not be accepted by everyone, having to think about it and consider it was, and is, stressful and contributes to Black fatigue.
Mary-Frances Winters, *Black Fatigue: How Racism Erodes the Mind, Body and Spirit*[2]

I remember the first time I heard the white world call me "nigger" … It was 1965. I was seven years old … I had never heard the word … I lived in a world where Blackness poured over us like warm molasses and filled our ears with affection, where Black love didn't have to be spoken to be felt. But I knew when the spell was broken and we were hurled into a parallel universe of quiet hate … I was a seven-year-old child feeling the weight of the white world's hate crushing my precious soul. There was no Benjamin Spock to explain the trauma I endured as a grade school victim of hate. There was no Jean Piaget to explain the impact of utter revulsion on my cognitive development.
Michael Eric Dyson, *Tears We Cannot Stop: A Sermon to White America*[3]

The first quotation above, from *Self-Esteem: A Classroom Affair*, a popular book published in 1978 by educational psychologists Michele and Craig Borba, reveals that experts now considered the transition from home to school as a critical juncture for self-esteem. Outside the family, the first "significant other" encountered by children was probably a teacher. As psychologist Ronald Geraty wrote in 1983: "In the charged atmosphere of the first day of school, one can begin to see the transfer of influence from parent to teacher. It is in school, with the new adult and with his peers, that the child first tests his view of himself as significant and competent."[4] However, Mary-Frances Winters and Michael Eric Dyson's recollections of their childhood memories remind us that classroom realities could look very different. Racism, from classmates and sometimes teachers, played a salient role in the forging of self-identity. It was that very moment, the first experience of white hostility, that African American mothers anxiously prepared for: the inner realization of white supremacy, racial discrimination, and a future life ahead of social disadvantage. Nothing in the Borbas' textbook prepared Black children for this in the way that both Winters's and Dyson's books later did.

In 2021, educationalist Stephen Vassalo described schools as places that help normalize "neoliberal selfhood." A neoliberal self is forged in schools organized around capitalist ethics such as individualism, efficiency, and production. Western schools inculcate qualities, attributes, and goals that broadly conform to capitalist agendas and economic values such as competitiveness. However, Vassalo believes that feelings of anxiety, stress, and isolation increase during the processes of orienting students in such ways. Applying psychology to education has had unforeseen consequences. In particular, he highlights school-based self-esteem initiatives as promoting a sense of self detached from social context, far too focused on inner experience, committed to self-mastery, and in need of constant expression. This type of education, described by its advocates as humanistic and positive, is potentially harmful for children's emotional wellbeing.[5]

Nonetheless, and particularly since the 1970s, anxieties about young people's emotions are usually addressed through school-based initiatives.[6] America's educational system has also proven useful for disseminating particular political visions. America's future was shaped in schools, but, from the 1970s, social conservatives were anxious to avoid leftwing, liberal values being inculcated in children early on in life, as they believed this would prove detrimental to national morality.[7] Educational psychologists began instead to imbue schools with a new sense of responsibility for supporting emotional growth, in addition to instructive learning. But, like much of American psychology, this turned out to be a fundamentally political endeavor. In this chapter, we will explore how and why schools became important spaces for transmitting the values of therapeutic culture, with self-esteem initiatives being top of the educational agenda. By the 1970s, the self-esteem concept had already been accepted so uncritically that educational psychologists were able to make a convincing case for pedagogic interventions based around the idea, hoping to replace older, often punitive, educational approaches oriented around teacher's authority rather than pupil's autonomy. In the so-called "affective classroom," self-expression was preferred to obedience.[8] However, from the outset, critics blamed this new emphasis on self-esteem for grade inflation, multiple valedictorians, and reducing education to a matter of making pupils feel good about themselves.

The nature of these educational interventions also reveals changes in the age groups of interest to psychologists. Freudians considered infanthood to be the most crucial time for psychosocial development, but late twentieth-century psychologists became interested too in the mid-childhood period, roughly 5–12-year-olds, arguing that this bore its own significance for personality development, and was similarly fraught with inner challenges. The Borbas argued that the older a child was, the harder it would be to change their psychological outlook: "early childhood education is, therefore, a crucial stage and an optimal time for affective learning."[9]

In this context, educational psychologists attempted to transform elementary schools into laboratories for optimizing personality development.[10] The idea of embedding self-esteem in school curricula was part of a broader pedagogical discussion about the most suitable educational technique. Were these traditional, subject-centered, rigid authoritarian approaches based on scholarly learning, or on modern, progressive, and child-centered research? Was knowledge for its own sake enough, or should learning focus also on socioemotional development?[11] Progressive forms of education cultivated individuality, free activity, learning from experience, acquiring personally relevant skills, and becoming acquainted with the world.[12]

However, some view the situation more critically. School curricula remained oriented around white people's values, despite desegregation. Skeptical educationalist Kenzo Bergeron warns that "self-esteem represents one of the most basic and pervasive mentalities ever to dominate the American psyche, one in which educators remain the primary moral agents and leaders in propagating its values." Bergeron views this negatively, for "through the cultural hegemony of educational socialization, individuals are conditioned to desire self-esteem. The obsessive focus on the individual self before all else has allowed for the reproduction of historic inequalities tied to race, sex and class."[13] Like other critics, Bergeron deplores the widespread application of the self-esteem concept in American education because it has so little to say about political, social, and moral matters. Entirely ignoring social injustice, individualistic education teaches children that self-help strategies, especially self-esteem building, provides enough fortitude to survive, later in life, in whatever sociocultural circumstances they end up facing.[14]

Self-esteem discourses still promised some kind of revolution, just not one intent on smashing the harmful forces of capitalism, racism, and social injustice. Instead, this was a revolt simply against the wrong ways of governing ourselves.[15] But regardless of their limitations, such approaches have permeated American schools since the 1970s. Self-esteem training, and other affective techniques, probably made children more inward-looking and self-reflective, but it also cultivated perceptions that all that's required to change life situations is to work on inner feelings, an approach that implicitly negated collective activism. School curricula treated self-esteem as a universal inner feeling and, in doing so, tacitly overlooked the specific needs and circumstances of students, particularly those from marginalized backgrounds.[16] Presumably, they had little to offer Winters or Dyson, whose memories of school racism left a lasting, lifelong negative impression on the psyche.

Educational psychology

The introduction of self-esteem initiatives in schools promised to improve emotional wellbeing, educational performance, and prospects in adulthood.[17] Particularly since the 1960s, psychologists published an overwhelming quantity of research linking education and self-esteem, arguing that high self-esteem was a deeply desirable trait to be cultivated in schools. The preferred type of pupil was self-expressive, self-managing, and enterprising (or, if one was feeling more cynical, self-absorbed, apolitical, and captivated by consumerism).[18] Humanistic educators drew extensively from Maslow and Rogers. One of these, Cecil H. Patterson, worked directly with Rogers on helping develop person-centered therapy. In 1973, he published *Humanistic Education*, a book that encouraged teachers to train pupils to be "sensitive, autonomous, thinking, humane individuals." Subject matter should be taught in a "human" way, using active learning and paying attention to the affective aspects of child development. These recommendations arose from Patterson's conviction that America's education system currently "suffocates the child's potential rather than releases it. Schools destroy the human spirit. They destroy the hearts and minds of children." For Patterson, schools dehumanized pupils, especially those who lagged behind educationally. Neither individuals nor society could advance by developing intellect alone, as children required socioemotional, as well as cognitive, education.[19]

Core pedagogic goals were reoriented toward growth, maturation, and behaviors.[20] Starting school offered the first significant opportunity for children outside the home to test their abilities to gain admiration and nurture self-esteem. However, humanistic psychologists insisted that traditional education threatened self-esteem. It was no longer good enough to sit pupils through endless examinations while watching their self-confidence evaporate. Education was too important to betray and harm young people in such ways. Low exam results stayed on pupils' CVs for life, diminishing opportunities in adulthood and fostering feelings of unworthiness even further.

Educational and child psychologists shared common approaches. Both had departed from researching only delinquent, psychologically maladjusted children to concern themselves with all children's emotional wellbeing. Both offered ostensibly helpful advice replete with ominous moral messages about the consequences of failing to nurture self-esteem. Both presented inculcating self-esteem as a moral duty for teachers and parents alike, for teachers, too, were now urged to devote considerable time and energy catering to children's emotional needs, while simultaneously working on themselves to optimize interpersonal interactions. All this was geared toward forging the idealized American personality, one brimming with self-esteem, confidence, and success. If children believed that they could succeed, then they would do so.[21] The new educational style was undoubtedly less punitive, maybe even fun, but did it realistically assess pupils' lives? Or did it instead discourage teachers from taking seriously important aspects of children's lives by stressing a naive optimism about inner improvement? And, even more problematically, was there an inherent danger in presenting low self-esteem as the *cause* of young people's various problems?

Earlier in the twentieth century, a small number of psychologists contemplated applying the self-esteem concept to education.[22] After the 1960s, more sustained pedagogic research was undertaken which typically criticized punitive education.[23] In 1969, one educationalist wrote:

> Students are also human beings. They have the same needs, the same urges, and are subject to the same principles of behavior as others. Schools, also, are not exempt from the operation of basic principles. In remembering these facts, we may substantially improve not only student performance but also the "level of gratification" in academic life.[24]

Expressing similar views in 1971, an English teacher based in Massachusetts, William P. Ferris, argued:

> Every year we turn out thousands and thousands of students for whom school is one of the bitterest experiences of their lives. Maybe if these students had been praised (yes, even when they wrote junk) instead of so often told and shown how stupid they were, there would be fewer psychological and sociological problems in our society.[25]

Ferris thought that too many pupils associated school with failure, and that teachers unwittingly reinforced this perception. "If teachers concentrated more on raising the self-esteem of their students," he argued, "they might have more of them to teach."[26] Progressive educationalists wanted teachers to stop conveying dry facts and instead promote "affective education," with the flourishing of self-esteem as a core goal.[27] In 1970,

Theodore and Nancy Sizer edited an enormously influential critique of current educational methods in which the cultural values previously being communicated through the American education system were condemned, and, instead, a new approach was propounded based upon pupils' rights. The volume led to discussions of self-esteem and student empowerment becoming de rigueur in the classroom.[28]

Educational psychologists warned that not all parents would have successfully nurtured self-esteem at home, making the teachers' task harder but all the more crucial.[29] Educational guides informed teachers that pupils arrived at the school gates with varying levels of self-esteem, depending on parental success, but it was never too late to act. In 1982, Robert R. Burns wrote: "Each pupil is already invisibly tagged, some enhancingly by a diet of nourishing interest and affection, and others crippled by a steady downpour of psychic blows from significant others denting, weakening and distorting their self-concepts. So children enter school with a self-concept already forming, but still susceptible to modification." Burns advised teachers against attempting to mold self-confident children, or treating them like "a malleable lump of clay." Instead, the self-esteem of each child first needed to be comprehensively assessed.[30]

An ever-growing number of educational texts on self-esteem appeared, including William Watson Purkey's *Self-Concept and School Achievement* (1970), which stressed the importance of how students saw themselves and their world. While acknowledging the complexities of educational failure, Purkey presented low self-esteem as a consistent running thread, a personal tragedy, and social waste. He warned that if negative self-images became embedded, teachers would struggle to dislodge them. Their prevention was vital to affective education and curriculum redesign. He identified six crucial factors: challenge, freedom, respect, warmth, control, and success. These could be attained by encouraging pupils to make more decisions, an alternative to older approaches that offered children little freedom of choice. To allow self-esteem to grow, pupils should be treated respectfully and never be made to feel embarrassed or humiliated. Purkey conclude: "Perhaps the single most important step that teachers can take in the classroom is to provide an educational atmosphere of success rather than failure."[31]

Humanistic psychology's influence continued to be felt in books such as *Self-Concept and the School Child* (1980), in which Robert Leonetti proposed a new form of constructive learning emanating from children's feelings of self-worth. Speaking metaphorically, Leonetti wrote:

> The development of self-confidence is like the growing, maturing oak tree. The tree's immediate growth potential and state of health are totally dependent upon its particular environment: climate, soil, sunlight, water. During a drought, its development will be retarded. If struck by lightning, it could grow crooked; if hit by disease it could wither and die. But no matter what, it continually strives toward growth and maturity. Its most intrinsic notion is to enhance itself, to actualize and grow into a strong and healthy oak tree.[32]

In common with other humanistic psychologists, Leonetti presumed that all children subconsciously wanted to enhance themselves and transform into the best person they could possibly become. For Leonetti, "the ultimate goal is to contribute to the development of the self-actualized person, one who is self-fulfilled, who is becoming all that he or she is capable of being." He recommended setting tasks that children could almost certainly perform successfully, starting off small before getting increasingly ambitious. From these successes, children's confidence would grow daily. Leonetti wanted children to enjoy some success each and every day, and advocated an intensively personalized education that paid attention to each pupil's differences while remaining focused on their personal strengths. All communication between teacher and pupil should be tailored personally and packed full with positive feedback.[33]

But the new pedagogic approaches contained various problems. How would children learn to manage failure? Did they risk developing an unrealistic sense of their own abilities? How would they cope upon leaving school and facing harsh adult realities? The transition away from punitive education was undoubtedly welcome, but were environments of constant, even unwarranted, praise really suitable? Moreover, was encouraging positive thought as a solution to all problems really a realistic start in life? Seeking solutions within one's self, and one's mind, was a superficial strategy accompanied by a high risk of proving unhelpful later in life when confronted with emotional environments strewn with unemployment, poverty, racism, and social isolation. But none of these flaws deterred self-esteem advocates from developing self-esteem-based pedagogies that fitted well into therapeutic culture and reflected democratic, conservative, white-oriented culture values.

Self-enhancing schools

Educators devised a seemingly endless stream of initiatives based on affective education, despite its many potential flaws. In the 1960s, the US Head Start program formed part of Lyndon B. Johnson's War on Poverty campaign. Head Start was a pre-school program enriched by social services aimed at economically disadvantaged children. Edward Zigler, a prominent psychologist, was actively involved. Initially, the scheme was well received, but it caused debate among psychologists, primarily about which was the most important lifecycle stage for nurturing emotional wellbeing. Zigler placed great emphasis on producing "socially competent human beings" through both cognitive and emotion-based learning, and developing a positive self-image. He claimed that 95 percent of parents enthusiastically supported the scheme. By the 1970s, he hoped that his scheme would be extended to all children.[34]

Some grassroots initiatives also surfaced. In 1966, Project Self-Esteem was established in Pittsburgh. Initially, the project involved engaging 250 Black children in Black music, art, and dance. The initiative was developed by African American communities which, as outlined above, were then harnessing the self-esteem concept for collective improvement. The scheme introduced children to Black American role models in American society and African heritage, offering information that was absent in the white-dominated education system.[35] However, the type of self-esteem education that extended across the 1970s education system abandoned any interest in societal matters such as racism. A later initiative, also named Project Self-Esteem, boasted an advisory board of white self-help experts: Nathaniel Branden, Dorothy Corkville Briggs, and Jack Canfield. The pilot program began in California in 1978. In 1985, Project Self-Esteem was awarded the Disneyland Community Service Award for its outstanding contribution to youth. Aimed at grades two to six, the program involved teachers and parents directing lessons on topics that included realizing one's uniqueness, compliments, feelings, friendship, social skills, conformity, and the dangers of alcohol and drug misuse.[36]

Such schemes appealed strongly to American values of individualism and one-size-fits-all approaches to emotional wellbeing that approached childhood self-esteem in generalized ways, paying little attention to more specific needs. Occasional concern was expressed about how African American children were faring in educational settings.[37] In 1973, it was proposed that African American children should learn to play the ukulele to boost their self-esteem.[38] But presumably much more was needed to tackle the problems stemming from racism. As the general focus of self-esteem advice moved rapidly away from the socially marginalized to instead helping all children grow into emotionally healthy citizens, into the types of people glowing with self-confidence and success, self-esteem's radical potential fell by the wayside.

Like initiatives aimed at family life, efforts were also made to reshape American school curricula around positive emotions and interpersonal interactions. Schools were to transform into democratic institutions, with authority shared between pupils and teachers. Ideally, children would enter nursery school and be greeted by a warm, caring environment. Teachers were provided with new instructions: don't do too much for the children; don't rush to help them put on their socks and shoes – let them develop the confidence to do this for themselves; don't scold the children – let them learn from their mistakes; don't belittle misbehaving children in front of others – this destroys their self-esteem; bathe them in unconditional positive regard.

As one 1975 book for teachers explained: "This kind of fundamental acceptance and approval of each child is not contingent upon his [the pupil] meeting the teacher's expectations of what he should be but simply depends on his being alive, being a child, and being in her group."[39] The key message was to deliver praise, recognition, and genuine respect. Educational books aimed at teachers, including Michele and Craig Borba's widely read *Self-Esteem: A Classroom Affair*, advocated creating enhancing pedagogic environments that would stimulate emotional growth and positive self-image. Strategies included asking children to say something positive if they were overheard uttering something negative, providing a warm welcome letter, celebrating birthdays, developing All-About-Me strategies, hosting self-awareness and friendship circles, keeping realistic progress records, and inclusively involving parents.[40] Classes of a more scholarly nature could be redesigned to accommodate both instructive and affective learning. English literature teachers might use reading materials in which authors expressed positive ideas about themselves. Science teachers could instruct on how to practically apply science to everyday life. Math teachers could emphasize practical skills such as budgeting.[41] In 1984, James Beane and Richard Lipka, authors of various self-esteem guides for the classroom, suggested that pupils could learn about the American Revolution not by memorizing dates, but instead by using the historical event as a springboard for reflecting on conflict with peers and family members.[42] All of this would contribute to the self-enhancing school, an educational site that was humanistic, self-directed, encouraging, full of positive expectations, and life-centered.

Nonetheless, danger existed in encouraging pupils to feel and comprehend such a limited range of emotions. Negative feelings were to be suppressed and dispelled. Pupils were actively discouraged from exploring literature that dwelt on despair, sadness, fear, or loneliness, even though these feelings were so common in adult life and, indeed, in the structure of capitalist modernity itself. The new educational style saw little need to ponder over tragedies in case this increased exposure to negative emotions. After all, the inherent sadness of tragedy was unlikely to produce democratic, happy, self-confident citizens. While spreading happiness and positivity might not seem like such a bad thing, clear issues existed in instructing pupils early in life to resolve life's problems from within, rather than without. If pupils simply imagined they were clever and likeable, happiness would ensue, regardless of broader life circumstances.

Guides such as *100 Ways to Enhance Self-Concept in the Classroom* (1976) provided practical exercises to be used by both teachers and parents. Typically, these involved students publicly sharing accomplishments, achievements, and success, discussing things they were proud of, sharing successes, learning how to accept their bodies as they grow, replacing negative language with positive.[43] But the emotional world and experiences of children were undoubtedly more complex, and far more fragile and fraught, than positive thinking approaches acknowledged. Humanistic psychologists drew from an almost caricatured, homogenous image of children whom they presumed were constantly striving to feel happy. Such imagery presumably failed to resonate with many pupils, especially those raised in conditions such as racism, poverty, and social disadvantage, whose problems were unlikely to be resolved by learning to play a ukulele or avoiding literary works that exuded too much sadness.

Redesigning teachers and curricula

Teachers were urged to reinvent themselves as emotionally attentive educators. Similar to parental guidance, self-help literature warned them of the severe consequences of neglecting children's emotional growth. As Purkey explained in 1978:

> Children need invitations the way flowers need sunshine. When they are treated with indifference, they are likely to become indifferent to themselves and to school. They begin to say to themselves "Give up. No-one cares about your small victories." One angry student bent on destruction can vandalize an entire school, just as one frustrated person with a rifle can redirect the course of human history.[44]

To achieve success, teachers needed to work on their own self-image too. Those lacking self-confidence could hardly command much respect from either colleagues or pupils. Teacher effectiveness emerged as a new goal in educational psychology.[45]

Like parents, teachers were spurred on by a rhetoric of guilt and fear. School counselor Verne Faust presented low self-esteem as a potentially explosive force. Low self-esteem, he warned, made pupils feel vulnerable and unsafe, caused anxiety and fear, and ultimately produced angry, enraged feelings toward teachers and parents. Youngsters who concealed this anger built up an inner resentment that could detonate unexpectedly in the classroom at any given moment.[46] Fortunately, all this could apparently be avoided simply by inculcating self-esteem and positive thinking. Similarly, Beane and Lipka warned that drug use, eating disorders, teenage pregnancy, crime, and suicide were commonly disregarded as episodes of youthful rebellion. They added: "How sad and misguided a view. Instead, these are symptoms of loss of self-worth, of loss of social place, of unclear values."[47] Their proposals included involving pupils in decision-making, developing cooperative governance systems, accepting diversity, and fostering self-esteem among both teachers and pupils.

The new task of teachers was to "invite" positive emotional reactions in pupils. Children who sensed that teachers saw them as trouble-makers or failures would internalize that perception and misbehave accordingly, but if teachers *believed* that students were able, valuable, and self-directing, the opposite happened. As Purkey elaborated: "Like a sculptor who envisions something in a block of marble that others cannot see, the invitational teacher sees possibilities in students that others miss." In that sense, both learning and self-esteem were dynamic, not static. Children were not born with fixed levels of skill, intelligence, and self-esteem; these needed to be nurtured. The direction children took was shaped heavily by the beliefs held by teachers about them.[48]

Suggestions were even made about architecturally restructuring schools around self-esteem principles. In 1983, one psychologist outlined the ideal school environment for nurturing self-esteem. It would have a relatively small number of pupils and would include initiatives such as pupil or artist of the week; it would consist of a single open classroom with separate areas for math, reading, art, science, and other activities; there would be no assigned desks. The teaching program would be flexible, individualized, and student-directed, although teacher-planned, and parents would be expected to get involved. There would be an emphasis on life skills such as self-worth, confidence, and motivation, and each child would be allowed to develop at his or her own pace. Children of different ages would interact, learning from one another.[49]

While it was clearly not feasible to rebuild all schools to accommodate the demands of self-esteem enthusiasts, many teachers developed practical initiatives. In 1992, the Mamaroneck Avenue School in White Plains, New York, introduced a motivational program that rewarded students for good behavior such as avoiding playground conflict. Students earned pencils, buttons, and points in a system that ultimately permitted a child to act as principal for a day. At Springhurst Elementary School, also in New York, teachers introduced a "child of the week" scheme. The lucky child got the opportunity to bring in meaningful objects from home and talk about their special talents.[50] Some of the new teaching strategies veered toward the nonorthodox. In 1989, pupils at a junior high school in New York were asked to stand in a semi-circle, shake their hands and bodies, and repeat strange phrases. While this might easily have been mistaken for a mystic ritual, the session was led by two professional actresses and was intended as a warm-up exercise for a writing workshop that would raise self-esteem and encourage self-expression. Word for Word was a small Brooklyn-based theater company that spent much of 1989 running similar workshops for students and parents.[51]

It seems clear that schools were encouraged to be oriented around emotional, as well as scholarly, nourishment. From a more critical perspective, it could be argued that the pendulum ended up swinging too far the other way. Schools, it was hoped, would transform into therapeutic landscapes intent on counseling and caring for children. Indeed, in 1982, one psychologist specifically stated that "the psychological events taking place within the classroom resemble therapy."[52] Yet skeptics insisted that this denied children opportunities to develop the real self-confidence that came only from genuine success, not constant praise. As education professor, Maureen Stout, argued in 2000:

> Now while I'm just as in favor of self-esteem as the next girl, I had always labored under the assumption that feeling good about oneself, the essence of self-esteem, was, or should be, the consequence of hard work, achievement, learning from one's mistakes, giving to others, trying to be a better person, or other similar endeavors. It had never occurred to me that I could bypass all that and just wake up and decide unilaterally (or with the help of my teachers) that I was a superb human being. I didn't realize that just by being in the world I was special, special enough to deserve grades, respect, and opportunities without ever having to earn them. But this is the dominant view among not only professors of education, but their students, our children's future teachers.[53]

Stout lamented how far self-esteem has permeated American schooling. Evaluation has become all about "making the kids feel good" while deceiving them about their true abilities (or lack of). For Stout, self-esteem initiatives have transformed the mission of public schools to serve individuals at the cost of quality, standards, and the public interest "without anyone getting wise to it."[54]

Self-esteem and Black children

Nowhere is the effect of white supremacy more pervasive and more debilitating than in the American school. Whether it takes the form of textbooks which promulgate white supremacy by excluding the lives and accomplishment of Blacks and other minorities, whether it takes the form of white teachers who have double standards of expectation, reward, and punishment, or whether it takes the form of self-hating Black teachers who despise Black children, white racism has poisoned the American school. White supremacy has left many Black teachers and white teachers paralyzed in its wake, and it has been most deadly when they are unaware of their social sickness."[55]

This quote, from Donald H. Smith in 1972, suggests a very different educational story for African American pupils. Still clinging on to the hopes of the late 1960s Black Power movement, Smith believed that Black children were taking hold of their destinies in schools by openly rebelling against a discriminatory education that "treated them as a sub-human species," for "these young people are determined that they will be respected, that they will be taught, that they will have access to the same opportunities available to whites. Whether the American schools recognize it or not, their Black pupils are in revolt."[56]

Smith was overly optimistic. From the 1960s, conferences and books on "negro self-concept" became slightly more sensitive to the unique problems facing Black children in schools.[57] A racially appropriate education was one of Black Power's key goals. However, by the 1970s community control of Black schools had proven largely unsuccessful. Black children still attended the worst schools; teachers in desegregated schools had few hopes for pupils of color. This was far from conducive to building Black pride and self-respect.[58] Some schools, usually under pressure from community groups, ran courses in Black history, hired a few more Black teachers, or drew attention to some Black heroes. However, the fundamental problem remained: a white-oriented education. In many desegregated classrooms, the Black child's "significant other," their teacher, would probably be a white American. Their pedagogical material, for the most part, still socialized Black children into a racist society that actively devalued them.[59] As another African American educationalist, James A. Banks, described in 1972:

> We have loosely and hurriedly put together a few courses on Black history, hired Black consultants to make brief visits to our schools, purchased a few multiethnic textbooks sprinkled with selected Black heroes, and placed a few

more Black teachers and administrators in our schools. We have done most of these things after a violent racial outbreak or a racial assassination. We usually retreat back into inaction after the flames have cooled.[60]

Whereas Black activists had hoped that educational equality would open up opportunities for success later in life, many studies from the 1970s argued that desegregation was actually worsening prospects for many Black children. Some educationalists hoped that the "open school environment," ones that supported affective education, would "override" socioeconomic status, allowing Black children to entertain positive hopes for the future.[61] In reality, new pedagogical approaches simply ignored ongoing problems pertaining to racism and lower socioeconomic status that were hardly likely to go away just because one's self-image improved at school.

Published in 1970, *The Bluest Eye* was Toni Morrison's first novel. Set in Lorain, Ohio, it told the story of Pecola, a young school-age African American girl who grew up shortly after the Great Depression. Pecola is regarded by most others as ugly and, as a result, develops an inferiority complex. She comes to desire the blue eyes that she equates with whiteness. The novel also covered the controversial issues of incest and child molestation. Morrison sought to express the psychological damage caused by racism among the age groups covered in this chapter. In part, *The Bluest Eye* was based on the author's own feelings of low self-esteem while young, her feelings of ugliness, caused largely by the implicit assumption that Black was far from beautiful. Pecola's characterization demonstrated the negative impacts of racism on self-confidence and self-worth at this formative school age. Published at a time when Black authors were engaged in writing powerful, aggressive fiction with racially uplifting themes, *The Bluest Eye* was a potent reminder that there existed another side of the coin. Internalized racism was a major theme of this novel, in line with contemporary interests in the psychological. Morrison later explained that the novel's origin lay in a conversation she had once had with a female friend when starting elementary school who told her that she wanted blue eyes. As Morrison explained, "Until that moment, I had seen the pretty, the lovely, the nice, the ugly, and although I had certainly used the word 'beautiful,' I had never experienced its shock." Morrison thought that implicit in the girl's desire was racial self-loathing and, "how something as grotesque as the demonization of an entire race could take root inside the most delicate member of society: a child, the most vulnerable member: a female."[62]

Yet parts of Morrison's novel subverted the idea of self-hatred. In a chapter early in the book, Claudia, a young Black girl who is the main narrator of the novel, tells of being given a big, blue-eyed baby doll. Instead of desiring the blue eyes, or viewing the doll as more pleasant looking than a colored doll, she described being physically revolted by the white doll, as being scared of its "round moronic eye, the pancake face and orange-worms hair." She had only one desire: to dismember the doll. "I destroyed white baby dolls. But the dismembering of dolls was not the true horror. The horrifying thing was the transference of the same impulses to little white girls. The indifference with which I could have axed them was shaken only by my desire to do so."[63] Eventually, she overcame her subconscious hatred of whiteness. She went through stages of sadism, fabricated hatred, to fraudulent love before coming to worship white celebrities such as Shirley Temple.[64] She internally succumbed to the messages being transmitted by the dominant white culture through institutions such as schools.

School desegregation was symbolically important, but many Black children had a worse time in mixed schools than in segregated ones, and Morrison was presumably among them. Although the idea of desegregation was undoubtedly well-meaning and symbolic, it was implemented in ways that devastated Black communities. It destroyed Black schools, reduced the number of Black principals and communities, and brought many Black children into contact with racist pupils and teachers.[65] Much depended upon the extent and nature of racism experienced, conscious or unconscious, both from pupils and teachers. In 1967, Robert Coles wrote of encountering some Southern teachers interested in teaching desegregated classes, but a majority who refused. As he wrote: "Their superiors did not wish to enforce a collusion of reluctant, angry or fearful teachers with nervous children, whites for the first time with negroes, as well as negroes with whites."[66] Winters, cited at the start of this chapter, recalled her first kindergarten experience of racism in 1956:

> One day Bobby (not his real name), a freckle-face white boy, called Karen and me the "n" word. We were not exactly sure what it meant, but we knew

it wasn't nice, so we started crying. The teacher came to our rescue and inquired as to why we were crying. After we told her, she called Bobby into the coatroom and told him that his red hair was ugly, and his freckles were too. While I am not sure a child psychologist would have concurred with the teacher's approach, it worked for us because Bobby was crying now too.[67]

In the North, whites generally supported civil rights goals until the problem moved into their own back yards (i.e. through migration), at which point their enthusiasm for integration faltered. There, white and nonwhite children largely attended separate schools controlled by a white leadership with little interest in ensuring educational excellence for pupils of color.[68] Even white teachers who considered themselves nonracist wrongly presumed that Black pupils underachieved. Too often, such harmful expectations created a self-fulfilling prophecy.[69] Even in schools with Black principals and teachers, state rules shaped curricula, limiting the choice of books and class content. Even schools that appeared "all-Black" still endured the underpinning influence of white power. Some considered the American school system unequipped and unprepared for desegregation. Others considered it, for the most part, simply unwilling to partake. Whites simply didn't want too many educated Blacks. Studies of desegregation usually found that Black children had more positive experiences in all-Black schools than in environments rife with racism where Black history and culture were handled insensitively. Black parents could exert little influence or leadership in many white-dominated desegregated schools. For Black pupils, integration did not provide a culturally salient education.[70]

By the 1980s, African American critic Derrick Bell concluded that civil rights and school desegregation had actually ended up serving white interests. Resistance had been fierce. Desegregation had helped schools attract funding for training, salaries, research, development, and new school construction. However, many school officials paid only lip service to redressing Black people's educational grievances.[71] Despite the ruling in *Brown v. Board of Education*, very little school desegregation actually took place in the late 1950s and 1960s. In the 1963–64 school year, only 1.17 percent of Black pupils attended school with white pupils. It was only the fear of losing funding that motivated many schools to comply. Reflecting upon his legal career in 2004, Bell recalled that he had once viewed the *Brown* decision as the "Holy Grail of racial justice." He joined a law faculty in 1969, by which point his hopes of schools complying with *Brown* through court orders had waned. By the mid-1970s, he doubted whether integrationist approaches could ever bring better schooling for Black children. Perhaps educational equity, not integrated idealism, was the solution.[72]

Even Black parents disproved. At least in a segregated system, their children were less exposed to racism, closer to home, and more likely to encounter curricula attuned to their needs. In contrast, integrated schools simply recited a white-focused curriculum, while too many teachers treated Black children as barely tolerated guests. By the 1980s, mandatory integration was falling out of fashion. Focusing on racial balance now seemed counterproductive.[73] The Reagan administration of the 1980s accelerated this process. As part of a more general roll-back of civil rights progress, Reagan oversaw new policies of dismantling the practice of bussing – whereby Black children were transported to schools outside their neighborhoods – and supporting ongoing school segregation, in addition to implementing significant funding cuts that impacted particularly badly on schools attended by African Americans, and which also led to a substantial decline in African American enrolment in university programs.[74] It comes as little surprise that references to African American life rarely featured in the educational strategies outlined throughout this chapter. In many schools, these students simply weren't welcome.

Black students were cognizant of the potential for even friendly white teachers to hold underlying discriminatory attitudes. As bestselling author Austin Channing Brown recalled about her own experiences:

> It was the first time I saw beyond my own perception of the racial harmony at my school. I was grateful that I didn't have to deal with overt acts of racism but was it better to know that teachers silently believed I would be a nuisance unless I proved otherwise? How could I know if beneath other amiable interactions, the stereotypes and biases of those in power were operating against students who looked like me?[75]

Brown elaborated by recalling how she then developed an antagonistic attitude toward her white-based education after first becoming aware of it. As far as possible, she opted to write about Black authors and Black history. She pondered intensely over whether to write on paper that Columbus "discovered" America. She preferred discussing Malcolm X over Mark Twain. But Brown was making a precarious choice: "I could choose the better grade or I could choose to affirm Blackness. It's a decision many students of color have to make."[76]

In American schools, Black children were expected to perform as well as white pupils despite the curriculum, texts, and teaching approaches being designed with whites in mind.[77] In 1977, one educationalist, Frederick D. Harper, criticized school curricula for being ethnocentric and class-centered, which kept "the disadvantaged Black child in a state of powerlessness, frustration, alienation, frustration, failure, low self-esteem, and a state of lack of control over his life." He added: "The disadvantaged Black student's self-esteem is further threatened by courses that reflect white, Anglo-Saxon, Protestant culture and books that reflect heroes that are not of his racial origin, or cultures that are alien to his poverty culture of the ghetto or his Black rural culture of the south."[78] At worst, school became a psychological nightmare for such children who, bored of their alienation, developed their own curriculum of smoking, rapping, card playing, and vandalism. Like many of his contemporaries, Harper favored a curriculum based around psychological concepts such as self-esteem, but thought it important, within this, to focus on developing proud, talented Black youth taught by committed school personnel who valued education for Blacks and whites alike, and genuinely wanted to help Black youths internalize a positive self-image.[79]

In *Raising Black Children*, Alvin Pouissant and James P. Comer were unafraid to publish questions that were mostly ignored in color-blind advice books, such as:

> My nine-year-old told me that his white teacher slapped him and called him a nigger. After I got to school and made a screaming fool of myself, he told me it was not true. What was going on?
>
> ...
>
> Some of the children in my first-grade class call each other names when they are angry – Black nigger, Black pig, and so on. How should I handle this?[80]

The authors believed that by around age 7 or 8, almost all Black children understood racial difference and social exclusion. Some became angry and hostile when confronted with an expectation that they should feel inferior, rejected, and insecure.[81]

To provide yet another example, in 1984, educational consultant Jawanza Kunjufu recognized the need for African Americans to be educated with frames of reference consistent with their culture. According to Kunjufu: "The liberation of African Americans is dependent upon an *effective* education. We emphasize effective because so much of our education has been virtually useless in accomplishing the objective of liberation." Kunjufu believed that Black children were not being "mis-educated": they were being "de-educated." By this, he meant that they were being systematically excluded from the educational system, or even destroyed within it.[82] (Kunjufu also believed that an instructive, psychologically oriented school system was inadequate even for white pupils, but doubly so for Blacks.) Kunjufu revealed that 42 percent of Black youths aged 17 could not read beyond a 6th-grade reading level; Black high school drop-out rates were as high as 49.6 percent; and Black children comprised 17 percent of the national school population, but made up 41 percent of the so-called "Educable Mentally Retarded" students.[83]

Kunjufu argued that African Americans had yet to develop a positive historical image that would enable their children to adopt a positive self-image. Indeed, he described the current system as an "American colonial education system," adding that "its claims of individual freedom, cultural pluralism and world democratization obscure its ideologies of elitism, cultural monism and world Americanization, and Black children are the acculturated victims."[84] In this account, Black children were being misleadingly educated to believe in their individual freedom, and to take on their own responsibility for success or failure in a society that persistently refused to grant them full socioeconomic and political equality.

Worsening the problem still further, many Black parents saw little need for Black and white culture to be taught alongside each other, still viewing education as "the three R's." Accordingly, Kunjufu urged them to reflect upon how their children might fruitfully accumulate self-esteem in a more appropriate education setting. He wanted African Americans to think about education as more than preparation for finding a job later in life. Ideally, it should instill in children a strong sense of identity, and also which forces oppressed them. For Kunjufu, education was about more than reading, writing, and arithmetic; it also needed to include self-esteem and values from an African perspective. He argued that developing a positive self-image in Black children was the parents' responsibility, with teachers providing supplemental nurturance. Only if parents had woefully failed should the school make an extra effort to develop positive self-images.[85]

Here, we can see that Kunjufu broadly agrees with psychologically driven pedagogical approaches, while acknowledging that these would benefit from having a stronger, more targeted Black dimension, ideally one that was more than just an afterthought. He insightfully recognized that Black children were notable for their absence in much research on child development and educational psychology. From the late 1970s onward, a small but visible number of African American researchers took it upon themselves to better understand, and draw pedagogical attention to, the specific needs of Black children. Janice E. Hale was among these. She advised Black children against trying to compete in "white" areas of achievement and to focus instead on areas of education in which they seemed naturally proficient (in her view, music was among these) or in which parents wanted children to succeed (e.g. mathematics). Hale identified in the American school system pedagogical processes designed for "Anglo-Saxon, middle-class children" that too often diagnosed under-performing Black children as hyperactive or mentally retarded when they struggled to excel.[86]

In an earlier chapter, I suggested that lower-class Black parents reared children for survival in a hostile, white-dominated society. Hale adopted a similar attitude toward schools, developing "education for survival" strategies. While she considered it important for education to celebrate African heritage and achievements, she thought it was also necessary to imbue Black children with skills needed to survive in America. She argued that Black children grew up in a distinct culture, and that educational psychologists needed to take this precept seriously.[87] Moreover, the self-esteem needs of Black children differed vastly from those of whites who, as a group, didn't face the sheer levels of hostility, discrimination, and social disadvantage faced by Blacks.

Evidently, the American education system was a key site in which the idea of self-esteem was nurtured and popularized, especially from the 1970s onward. Educational psychologists tended to decompartmentalize life's problems and to see their resolution as resting in simple acts of emotional change, most notably positive thinking. Here, we can see the ongoing influence of humanistic psychologies and their hopes for inner change, albeit shorn of the counterculture's interests in broader sociopolitical change. Of course, inevitable problems arose with only viewing life's problems from one angle (the inner and emotional) without considering how broader socioeconomic and political forces shape our inner feelings by creating particular "emotional environments." We see once again tendencies to ignore all the strident personal and social work undertaken by Black and LGBT activism with regards to ideas about self-esteem (and related terms such as self-respect), a subtle re-enforcement of white cultural superiority that operated simply by ignoring people and personality types that didn't fit into the idealized American norm. In the next chapter, our discussion moves to teenagers, a social group targeted especially thoroughly by self-esteem initiatives and which provided a focus of discussion in postwar America for those concerned about the future direction of American society.

SIX

Teenagers

Until the twentieth century, we didn't view adolescence as such a particularly meaningful transitory period as we do today. The "teenager" was a new twentieth-century invention, and one who caused considerable concern. After World War II, a vibrant youth culture emerged awash with new music styles, rebellion, and nonconformist ideals. Parents and teachers felt increasingly unable to impress traditional values on the young, a group that had developed its own culture. Changes in youthful activity culminated in the 1960s counterculture. Around the same time, doctors and psychologists developed specialist interests in teenagers, and even established a new subdiscipline: adolescent medicine. The emergence of an independent youth culture caused strain within families as parents struggled to adapt to their teenagers' new-found autonomy. The conservative right felt threatened by leftwing culture, its ideals and its attacks on the family. In response, psychologists and doctors actively interceded in this growing gap between generations.[1]

Dominick Cavallo argues that permissive, emotionally bonding child-raising techniques, popular from the 1950s, promoted self-reliance, self-esteem, and competitiveness. In turn, the generation that came of age in the 1960s had enough self-confidence to be opinionated, rebellious, and, indeed, to kickstart a counterculture.[2] Jean Twenge adds that the confidence of American teenagers continued to rise well into the 2000s. Most teenagers became reasonably satisfied with their bodies, appearance, intelligence, and capabilities. The intent focus on self-confidence in schools aided these positive feelings.

So far, so good. However, Twenge believes that twenty-first-century teenagers are in fact *overconfident*, to the point of narcissism. Decades of boosting self-esteem in families and schools created a deep-rooted sense of entitlement. For Twenge, the "GenME" generation holds overly high expectations that, if they follow their dreams, they will be successful and, preferably, famous. She argues that this creates widespread disappointment. The inevitable dissatisfaction of life increases levels of teenage anxiety, depression, suicide, and loneliness and young adults are frequently disheartened by the realities of life. Expectations of securing good jobs and nice housing are frequently crushed. Years of self-esteem instruction and of telling children that they are special and can achieve anything left people feeling confused and hurt by life's harsh realities upon entering the adult world. As Twenge explains: "Our grandparents weren't lonely or obsessed with achieving fame and stardom ... we long for the social connections of past years, we enter a confusing world of too many choices, and we become depressed at younger and younger ages."[3]

Adolescence has essentially served as a testing ground for checking the success of emotional nurturance by parents and teachers. Postwar psychology supported forms of selfhood required of democratic citizens. Democracy depended upon maturity, mental stability, and freedom from the psychological forces that bred aggression in authoritarian cultures. The family, and its ability to emotionally support children, became central to psychological theory and practice. Families and schools were the basis of social democracy and incubators of citizenship and community values. However, psychologists often saw, or at least depicted, children as helpless and aggressive. This framing of childhood created a space for experts to help remake and sustain democracy.[4]

The teenage years offered the last opportunity for intervening in psychologically wayward childhoods and for instilling the values and mindsets appropriate for an ideal American adulthood. In the twentieth century, relatively little research was published on self-esteem in teenagers. Hence, much of this chapter relies upon what adults were saying about teenagers. Typically, experts dwelt on the negatives of teenage life and sought to salvage adolescents from alternative lifestyles, pregnancy, and clichés around "drugs, sex, and rock 'n' roll" lifestyles. Whereas psychological and self-help literature on younger children had focused on self-growth, discussion of teenage self-esteem dwelt mostly on the harmful consequences of underconfidence. But could it be that a relative lack of interest in "normal" happy teenagers created a distorted impression of adolescence that overemphasized deviance and antisocial behavior? I think so. It is certainly possible to interpret this genre of psychological writing as an anxious conversation with "the youth problem," revealing more about the concerns of the authors than about the young people themselves. Such literature articulated sociocultural concerns about teenagers that barely reflected the realities of youthful lives.

Of interest to this chapter is how teenage problems became reframed, managed, and (potentially) resolved first and foremost as a matter of low self-esteem. Rather than expanding much further on pedagogical approaches, already discussed in depth throughout the previous two chapters, here I focus instead on how low self-esteem was blamed from the 1970s onward for an all-encompassing range of social issues: occultism, drug abuse, juvenile crime, delinquency, sexual misconduct, teenage pregnancy. Once again, efforts were made to police these problems through individualized self-esteem initiatives bereft of broader sociopolitical context, which intervened obsessively at the myopic level of emotions, families, and interpersonal dynamics.

Under their ostensibly helpful appearance, self-esteem initiatives were fundamentally political. We see here the increasing convergence of self-help with the mounting influence of neoliberal and capitalist mentalities. Advice for teenagers strongly promoted social conformity and shoring up (ideally nuclear) families. It openly reacted against the counterculture that had, ironically, offered teenagers confidence and independence. Whereas counterculture activists had seen their rebelliousness as *expressions* of youthful self-confidence, psychologists and other critics turned this principle on its head and recast such behavior as problematic *symptoms* of low self-esteem. Only by reasserting the sociomoral importance of the nuclear family, and by encouraging conformity rather than rebellion, could adolescents regain a true, authentic sense of self-esteem. Predictably, this particular use of self-esteem theory ignored the emancipatory potential of self-esteem evident in the radical leftwing politics of the Black and LGBT rights movements, by transmitting the cultural values of white, middle-class America.

Pathological teenagers

Social psychologists considered the teenage years to be crucial for self-esteem but usually portrayed the emotional world of teenagers pessimistically. Foundational texts on adolescence included Granville Stanley Hall's 1904 study of adolescence. Now widely considered as the first "pioneer" of adolescent research, Hall famously characterized adolescence as a time of "storm and stress."[5] In the 1960s, Anna Freud depicted adolescence as a period of "developmental disturbance" during which teenagers came to terms with their sexual urges.[6] Contemporaneously, prominent German émigré psychoanalyst Erik Erikson coined the term "identity crisis" to describe the emotional upheaval of the teenage years.[7] Adolescence was presented as filled with physical and emotional hazards. Rather than enjoying the prime of life, teenagers, as portrayed by medical professionals, suffered widely from eating disorders, suicidal tendencies, self-harm, drug abuse, and smoking and alcohol dependency. Some psychologists ominously warned that self-destructive behavior was prevalent in one in five teenagers.[8]

Then again, adolescence could potentially be a positive experience. More optimistic interpretations pictured it as a phase of rapid self-growth in which youths distanced themselves from their parents to emerge as independent adults. Self-esteem was derived from personal achievements, although the pool of significant others was important and now encompassed peers and close friends as well as parents and teachers. Multiple opportunities existed to develop autonomy and freedom.[9] Nonetheless, the overarching message insisted that the path to self-confidence, autonomy, and adulthood was fraught with threats and obstacles. Teachers were rendered as "emotionally vulnerable," a relatively new phrase at the time.[10] Contradictorily, these newly independent youths still needed the protection of experts, parents, and teachers.[11] In relation to psychologies, adolescence involved an entire restructuring of the self.[12] Self-esteem fluctuated as teenage bodies and minds changed, but satisfaction with the new selves inevitably varied. As John Mack and Steven Ablon explained in 1983: "Self-esteem is continually being lost and restored in adolescence and adult life. Its reworking and revision, together with the maturation of the agencies responsible for its regulation, are the central tasks of human life. Of one thing we are certain: a forfeited, damaged or decreased self-esteem is an unacceptable condition."[13]

In 1965, Morris Rosenberg published the first scholarly monograph devoted entirely to teenage self-esteem in which he outlined his investigations into self-attitudes among more than 5,000 adolescents drawn from New York's high schools. Rosenberg was a social psychologist and sociologist working at the National Institute of Mental Health, and he identified teenagers as a group acutely concerned with self-image. After all, they lived through a life stage of major decisions (which occupation to choose?), burgeoning sexual awareness (am I physically attractive enough?), and physical change (do I like my new body?). Teenagers seemed unclear about their social identities, positions, and responsibilities, being not quite adult, not quite child. As their adolescent bodies were changing rapidly, doubts about self-image became particularly pronounced as teenagers came to terms with an entirely new physique.[14]

The *New York Times* reported positively on Rosenberg's book, hoping that those who procured high levels of self-esteem would become "a remarkably productive sort of citizen," a class officer or leader who would join clubs ranging from Spanish, to chess, to hiking. This was the sort of student who kept in touch with the news, eloquently discussed political affairs, enjoyed competition, and had confidence in their own ideas.[15] Evidently, the newspaper hoped that interventions in teenage self-esteem would produce an implicitly white, largely male, middle-class American citizen, an odd rejoinder to the contemporaneous emphasis on self-esteem and social justice prevalent in the civil rights movements.

Rosenberg was especially concerned with how self-esteem among teenagers dropped as they transitioned from primary to secondary education. Whether this happened due to changes in biology or to educational environment seemed unclear. In subsequent publications, Rosenberg elaborated on the precariousness of self-esteem at age 12 when self-image became destabilized, self-esteem lowered, and negative self-perceptions bubbled to the surface. Rosenberg identified this as a time when blossoming youths relinquished their inflated self-image of childhood and instead developed realistic self-perceptions.[16] The attention now being paid to teenage emotions illustrated another important cultural change. Since the nineteenth century, teenage delinquents, especially those from "broken homes" had typically been regarded as inherently immoral and sinful. More often than not, they frequently found their way into prison or some other kind of youth incarceration institution.[17] However, Rosenberg's approach reflected a shift toward considering teenagers, even delinquent ones, as emotionally vulnerable, whose problems could be managed in clinics rather than in prisons.[18]

Nonetheless, much self-esteem literature still depicted wayward teenagers as emotionally wounded, psychologically sick, and in need of protection from relentless inner turmoil. Teenage rebels were (sometimes) sympathetically recast as "disturbed adolescents" who were "acting-out" to express their feelings, conflicts, and emotional pathologies.[19] This was a more compassionate perspective than had been commonly presented in the recent past. The idea that teenage behavioral problems might be resolved clinically offered more hope than their being sent to the stigmatizing criminal justice system. That said, psychologists remained fixated over emotional pathologies among adolescents, a quite curious development given the general emphasis toward the positives in humanistic psychology. A few researchers investigated "normal," "healthy" adolescence, but a vast majority veered toward delinquency, antisocial tendencies, and emotional volatility. Adolescence remained caricatured, and set apart, as a time of affective turmoil and low self-esteem, despite compelling evidence of diversity in teenage experience other than "storm and stress" and specific needs among those from marginalized backgrounds.[20]

Psychologist Howard B. Kaplan linked low self-esteem to teenage delinquency. Kaplan was convinced that teenagers typically turned to crime to boost their flagging self-image. Nonetheless, like many of his contemporaries, he believed that humans were driven by a natural need to maximize positive self-attitudes. Teenagers viewed negatively by others felt emotionally distressed, and used the thrill of deviant behavior to bolster their self-image.[21] Kaplan's ideas were (and still are) heavily debated in academic circles.[22] At the time, other researchers thought that self-esteem declined only as a result of the intense guilt experienced after a teenager committed a crime.[23] What is clear is that the self-esteem concept was now being applied, in one form or another, to understand and police the "youth problem," opening up possibilities for therapeutically managing social problems. Ideas initially popular in the counterculture about self-growth and self-actualization were now being harnessed, somewhat ironically, to help restore family tranquility and social order. In that sense, the self-esteem concept was just as useful for supporting conservative, as well as emancipatory, politics. A clear tension had emerged in matters of self-help.

Since the nineteenth century, Americans had taken pride in their schools but attendance was intermittent for most. Only those expected to progress to college attended school on a regular basis. By the late twentieth century, school was where youngsters spent the majority of their time. By then, America had comprehensive, publicly funded, and widely available high schools open to all, and attendance became an expected part of the adolescent experience. In 1940, only 15 percent of 18–21-year-olds went to college; by the early 1960s, that had risen to 53 percent. Schooling became a standard part of the extended adolescence that prepared children for adulthood in America.[24]

Nonetheless, while it was clearly desirable for teenagers to gain independence, suspicion was cast on the interpersonal relations between peers that were taking place out of parental sight. Although healthy levels of self-esteem might provide teenagers with enough confidence to resist peer pressure, low levels of self-esteem made teenagers vulnerable to peer pressure and to being led into lifestyles involving drugs, drink, and crime. In environments unconducive to self-esteem, teenagers, it was feared, suffered from a spectrum of feelings and problems: anxiety, depression, suicide, underachievement, alienation, delinquency. Accordingly, writers such as James Battle warned teachers and parents in no uncertain terms that failing to nurture the young, emotionally and intellectually, could have dire, potentially fatal, consequences. Worse still, teenagers spread their emotional "illnesses" through their peer network, meaning problems such as drug misuse spread like an "infection."[25]

In reality, many people presumably still viewed rebellious youth as inherently "bad," even as evil. Nonetheless, a growing consensus saw troubled teenagers as emotionally scarred by harmful domestic and interpersonal environments over which they had little control. This new approach opened up therapeutic possibilities for amending negative emotions to help teenagers pursue a new life of happiness (or, if one was feeling more skeptical, encouraging conformity to socially acceptable behavioral norms). Some experts now surmised that low self-esteem was the very reason young people had joined leftwing subcultures in the 1960s. Only those who enjoyed high self-esteem had proven "well" enough to resist the counterculture and see the existing democratic social order in a positive light, as a force to be worked with rather than rallied against.[26] Here, we see a particularly striking example of how self-help and therapeutic culture often formed part of a conservative backlash against leftwing ideals that sought to impose conformity on teenage life.

Teenage Satanists with low self-esteem

Approaches that reduced teenage problems to a matter of self-esteem were hopelessly naive. Nonetheless, they reflected growing tendencies to locate causes of social problems in interpersonal relations and familial life and then offer assistance through therapeutic self-help practices, all in a self-enclosed vacuum from the outside world. By the 1980s, the boosting of adolescent self-esteem was truly thought to protect against a broad and despairingly ambitious spectrum of problems: cults, drugs, alcohol, pregnancy, radical politics, teenage rebellion, off-beat religious practices, witchcraft, Satanism, and so on. When choosing their preferred method of rebellion, teenagers seemed to have various options laid out before them. Kaplan thought that teenagers' personal life history determined their chosen path. Only by developing confidence could teenagers resist this plethora of threats to their wellbeing; only by experiencing familial love could they begin to participate in "responsible activities."[27]

In the 1980s and 1990s, Satanism caused a surprisingly huge moral panic.[28] Fears proliferated that teenagers across America were participating in devil worship and dark rituals. Experts speculated whether "the lure of the secret, powerful world of magic has an equally high appeal [as drugs] to youth with low self-esteem ... another path for dealing with life's problems and for expressing individual identity."[29] In 1990, the National Association of Secondary School Principals (NASSP) issued a "special bulletin" that warned parents and teachers to stay alert to signs of Satanism, such as the tabletop role-playing game Dungeons and Dragons, heavy metal concerts, and private parties rife with sex, alcohol, and drugs. While some of these activities might seem relatively innocent, warned the NASSP, they could can culminate in ritual-related crimes, trespassing, arson, grave robbery, cruelty to animals, kidnapping, rape, and even murder.[30]

Psychologists began to frame Satanism and occultism as an attempt to meet particular psychological needs, including an ingrained desire to acquire self-esteem. They conjectured that magic and the black arts offered a sense of control, power, dominance, and self-worth that helped underconfident teenagers feel better about themselves.[31] In 1990, the NASSP warned that Satanist teenagers were usually "very intelligent, but may be under-achievers with low self-esteem and a feeling of powerlessness," adding that they were typically white and middle class. Furthermore, "they are highly creative and curious. They have few friends and don't relate to peers well or fit into normal teenage social life. They may be alienated from the family's religion."[32] Similarly, in 1990 the International Cultic Studies Association (ICSA) described youths who joined Satanic cults as "alienated, troubled teens with low self-esteem who exhibit problems with aggressive behavior and/or suicidal tendencies, both of which can be aggravated by involvement in Satanism." The ICSA added that "teens who feel powerless, victimized and isolated may find that Satanism provides a sense of control, status and belonging. Intense group identity and bonding may result when a cult forms around a charismatic peer or adult who acts as high priest."[33] Training manuals offered to police forces by the US Department of Justice in the 1990s similarly described the profile of Satanists as intelligent but underachieving, creative, curious, often middle-to-upper class, but with low self-esteem and difficulties relating to peers.[34]

One example cited in the ICSA was that of Sean Sellers, convicted of murder in September 1986 at the age of 17. Sellers was a troubled, though academically gifted, teenager thought to have turned to Satanism and fantasy role-playing games for solace and power. With a group of friends, he formed a Satanist cult that performed blood-drinking rites. Sean's grades began to drop, he lost interest in sports, started abusing animals, and then acquired a heavy dependence on alcohol and drugs. In September 1985, he shot dead a convenience store clerk who had refused to sell him beer and, later, his mother and stepfather while they were asleep.[35] An unconvinced jury made little of Seller's plea of innocence (he denied responsibility on the basis of having been possessed by a demon), found him guilty of multiple homicides, and sentenced him to death.[36]

Cases such as this provoked fears that even academically bright youths (worse still, even middle-class ones!) might opt against conforming to salvage their flailing self-esteem, initiating a descent into a life of substance abuse, sexual misconduct, and ritualistic murder. No longer did the typical teenage delinquent come from a "broken" working-class home. Psychological dysfunction could potentially afflict any teenager, with little regard for class boundaries. All teenagers were now cast as vulnerable. Psychiatrist Saul Levine described the emotional experiences of joining a Satanic cult:

> During the period of ultimate commitment to the group, they are often alienated or estranged from their parents. They are suffused with self-righteousness and justification. They feel especially sanctified and enhanced. Not only are their earlier feelings of alienation, demoralization and low self-esteem overcome, but they feel "wonderful" in all respects. They experience happiness, clarity of thought, and high motivation.[37]

Once ensnared in such cults, teenagers were gradually coerced into dismembering human victims (often children), consuming their blood, and eating their flesh and vital organs. As Satanism promised power, dominance, and gratification, adolescents who felt alienated, alone, angry, and desperate were easily seduced. Participation in Satanic worship met their psychosocial needs. In such accounts, Satanism was remade as a problem of personality stemming from family conflict and alienation.[38]

But if Satanism was an inner, personal problem, and less of a social one, this should have raised questions about why so many emotionally disturbed and delinquent teenagers apparently felt such powerlessness and suffered from low self-image in the first place. Satanic teenagers gained little material benefit, unlike, say, youths who shoplifted. The benefits were purely emotional: increased feelings of power and self-worth. Indeed, therapeutic culture created an environment in which public attention now coalesced around psychosocial causes and implications. Before long, statements such as this were being made: "It would be useful for community agencies to develop youth programs aimed at enhancing the self-esteem of socially ostracized and alienated, middle-class adolescents."[39] Parents were equally urged to pay more attention to their children should problems develop. Typical messages included: "Don't panic about a few heavy metal albums." "Express an interest in heavy metal, even if you do not approve." "Examine your own behavior." Are you setting a good example as a successful role model?"[40] But here we see how far the pendulum could swing toward a focus on the white middle-class child in self-esteem discourse, and how estranged these concerns were from self-esteem's more radical roots in social justice concerns.

Self-esteem and teenage problems

The framing of adolescence as a time of burgeoning autonomy made the life phase appear as both welcoming and menacing. Teenagers needed enough emotional resources to be able to resist antisocial behavior, although many seemed not to possess these. Only adolescents with high self-esteem were likely to develop into the ideal democratic American citizen, one who abstained from drugs, alcohol, and pre-marital sex, who performed well at school, thought carefully about future careers, and aspired to raise a stable nuclear family. But the problem remained: many people lacked levels of high self-esteem. This could be ascribed to factors such as poverty, socioeconomic disadvantage, or discrimination based on race, gender, or class, but few people still thought this way. Even race was no longer a burning issue in the somewhat homogenous discussions about teenage self-esteem. Simply working on one's self-esteem was presented as enough to surmount life's barriers. Once this had been achieved, it might then be possible to find a compatible partner, purchase the finest cars, and buy a pretty-looking suburban house. Little wonder then, as Twenge believes, that teenagers grew up disappointed and dissatisfied.[41] For working on one's emotional life simply could never guarantee access to the materiality of postwar consumer life, particularly for teenagers living in inequitable environments.

To provide a few examples of how this worked in practice, the flourishing of self-help ideas from the 1970s brought new ideas concerning how drug addicts could be taught to help themselves and overcome their problems, albeit with some guidance.[42] Teenage drug use was reframed. No longer seen as a moral or personal failing, it was instead considered as a psychological effort to self-medicate negative emotions: shame, rage, loneliness, depression. Drug addiction was recast as a problem whose root cause was low self-esteem and deficient family upbringing. As was argued by Robert Millman and colleagues:

> The self-administration of psychoactive substances remains one of the few pleasurable options for many young people; it may be a predictable, reliable way to punctuate an otherwise unrewarding existence. The drug abuser obtains peer acceptance, a sense of control and a pharmacological effect that enhances self-esteem and relieves tension and anxiety.[43]

The psychological and personality factors underpinning drug use, including low self-esteem, had first been explored in the 1960s.[44] In subsequent decades, this idea really took off, and marijuana smokers were rethought of as people suffering from insecurity, self-rejection, and low self-esteem. They turned to marijuana to cope better with their psychological maladjustment. Of course, both low and high self-esteem individuals used marijuana, but the former had particular reasons for doing so. Because they found life so dreadfully unfulfilling, the boost and highs obtained from drugs were far more enjoyable and intense than those experienced by happier teenagers.[45]

Most teenagers knew about the risks involved in smoking, drinking, and taking drugs, but many of them indulged anyway. In light of that, it could be argued that such individuals were suffering from low self-esteem and were unable to resist peer pressure. If so, then the solution to teenage drug use rested, predictably, in improving self-confidence, self-control, and ability to cope with anxiety.[46] The relation between drugs and self-esteem was presented as cyclical. Low self-esteem led to drug use; a new identity as a drug user caused further self-esteem decline. Sometimes, the high produced by drugs might be the first true feeling of euphoria ever experienced, encouraging progressive use. The solution, other than keeping teenagers away from drugs, was to strengthen their self-image. But such approaches did little to address the multifaceted reasons why young people used drugs, or how drugs arrived in their communities. This was a Reaganite approach built on empowering people cost-effectively and which, not coincidentally, accompanied reduced funding for social programs.[47]

From the 1980s, school-based drugs programs incorporated self-esteem initiatives aimed at helping teenagers to make responsible decisions. Hence, the focus was on personal empowerment through building levels of self-confidence. Until then, teachers had warned of the physical and mental hazards of drug-taking, but many teenagers ignored these warnings and continued experimenting. As an alternative, self-esteem-based approaches aimed to strike at the perceived heart of the problem. Anti-drugs programs now featured "life education" aspects intended to promote responsible decision-making and, in turn, build self-esteem. The focus shifted toward developing the skills required to resist peer pressure and "just say no." Additionally, those already addicted were to be treated as individuals requiring rehabilitation, because punitive punishments could lower self-esteem further still. The general trend was to abandon scare tactics and help build school pupils' positive identities, underpinned by healthy self-esteem levels, in the hope of making drugs seem less attractive. Anti-drugs advice focused on values and individuality rather than simply reciting information on the dangers of drugs.[48]

One initiative was named Say It Straight (SIS). The idea was that students would learn straightforward communication skills and receive positive peer support, thereby boosting their self-esteem. Students filmed themselves acting in various interpersonal situations to explore how they felt and the effects their words had on others. As actors, students who usually exhibited negative personality traits could explore a wider repertoire of emotional behaviors such as caring. Providing factual information about alcohol or drugs was not a significant part of SIS training; rather, the focus was on skills and interpersonal behavior. It taught young people how to say no (while presuming that they had a tendency toward wellness, not destruction). Some schools using the program claimed to have seen rapidly lowering numbers of suspensions for alcohol and drug-related incidents. According to this line of thought, simply learning to communicate in positive ways helped prevent a plethora of social problems. As one enthusiast insisted:

> This training gives students the opportunity to discover their deepest wishes in difficult interpersonal situations and to develop skills necessary to implement them, thereby enhancing their self-esteem. They begin by exploring feelings and the reasons they have said yes in situations in which they very much wanted to say no. Invariably, they discover that they share fears of being rejected or not liked, or hurting a friend's feelings, of being embarrassed or not "cool." They discover that because of these fears they betray their deepest wishes.[49]

Teenage smoking, too, was recast as related to self-esteem levels. After all, why would anyone take to a poisonous, deadly habit if they loved and respected themselves?[50] As with anti-drug initiatives, most educational programs had initially disseminated ominous warnings about health consequences, but the strategies that replaced them focused also on emotions. From the 1980s, psychologists argued that smokers had weaker psychological resources than non-smokers, as was manifest in their lack of self-control. Such adolescents placed little value on their health due to their flailing self-esteem, and smoked to improve their woefully sad, self-deprecating mood. Worse still, teenage smokers' weaker psychological resources meant that they struggled to manage stress, rendering them even more vulnerable to smoking.[51]

Although teenage pregnancy rates actually fell in the 1970s, partly due to abortion legalization, teenage sexuality continued to cause concern.[52] In earlier decades, teenage pregnancies had been presumed to be accidental, or a result of seduction by an unscrupulous, usually older, man. New lines of thought emerged in the postwar period along the lines that some teenagers *chose* to become pregnant believing that a baby would provide the love and acceptance missing in their lives. In 1981, Lucie Rudd argued that:

> When the young girl has been lacking a feeling of being loved and cared for, and especially if she has been placed in foster homes or institutions, she will often develop the fantasy that a baby of her own will give her the love she has missed out on and will help her to live happily ever after. Little does she realize the amount of time and care that the baby will require ... A sense of belonging and approval in early childhood might prevent low self-esteem in early adolescent years.[53]

The list of character traits and problems associated with teenage pregnancies was both grim and exhaustive: "broken homes," lack of trust, drug and alcohol abuse, permissive sexual attitudes, early sexual activity, and, of course, low self-esteem scores in tests.[54] In 1974, psychologist Virginia Abernethy blamed poor parental upbringing for supplying the conditions of low self-esteem that encouraged risk-taking sexual activity. She insisted that birth-planning strategies should focus on psychological variables, since unwanted pregnancies were linked inextricably to familial pathology.[55]

Once again, a problem once considered sociomoral was recast in psychological terms, reinvented as an issue stemming from interpersonal relations and turned into an issue requiring mental health attention. As with Satanism, drugs, and smoking, the emphasis was on the emotional causes of teenage pregnancies: feelings of marginality, helplessness, and inferiority, a weak ego, lack of impulse control, inability to deter gratification.[56] Pregnant teenagers suffered a sense of ominous fatalism about their future prospects. Feelings of worthlessness led to decisions to get pregnant. Of course, such statements can be interpreted more as a commentary on middle-class sexual values and that group's presumptions about working-class teenage girls' bedroom behavior. Seemingly distraught about their lack of sexual morals, single teenager mothers threw themselves into parenthood. Teenage virgins apparently enjoyed far higher self-esteem levels than non-virgins, as did church-goers, non-smokers, and non-drug users. High self-esteem was thus closely associated with morality.[57] As with most other attempts to pin the low self-esteem label onto teenagers who were deemed problematic, researchers disagreed or struggled to find convincing empirical evidence about the precise links between teenage pregnancy and self-esteem.[58]

In 1991, social psychologist Roy F. Baumeister spoke of alcohol, drugs, and suicide as an attempt to escape the self, for "to escape the self is to free oneself of the struggle to maintain a certain image."[59] This idea of escapism was a common running thread on the discussions about Satanism, substance abuse, and teenage pregnancy. Yet it was not so much society that needed to be escaped from, at least according to the experts; rather, it was inner feelings of insecurity and low self-esteem. Of course, the problem here, as with many self-help strategies, was that teenage self-esteem was seen as distinct from socioeconomic, political, and cultural contexts. Hence, it was not the working-class teenage mother's lack of economic opportunities that encouraged her to become pregnant, or a lack of access to sexual education, but, instead, her woeful inner feelings. Sending her off on a self-esteem course might have been a quick-fix, but the adverse emotional environment that restricted opportunities, and perhaps also the availability of sexual knowledge, remained resolutely in place.

Self-help strategies encouraged teenage mothers, drugs users, and Satanists to adjust to their surrounding environments, no matter how adverse or dire these might be. It is notable, particularly in relation to Satanism, that these were problems perceived to have considerable potential to afflict middle-class white families. As in the previous two chapters, we can see how ideas about self-esteem rapidly permeated important aspects of life for young people, starting from motherhood, working through the school years, before culminating in teenage life. This chapter on teenagers demonstrates particularly clearly how all manner of social problems were blamed increasingly on negative emotions and mental states, with low self-esteem being the most prominent of these. In turn, the logical solution was to work on oneself and improve one's emotional state, ideally becoming happier and more self-confident. This process commenced while young with the help of parents and teachers, before reaching an age when responsibility for oneself was possible. Of course, there's nothing wrong with encouraging young people to feel happy, confident, and good about themselves, but the approach was myopic in the sense that it drew attention only to the inner aspects of life's difficulties, while obscuring the structural, social, and political. This was bad enough in itself even for the "normal" (i.e. white, middle-class) child, let alone for the Black or gay teenager, or teenagers from any other marginalized community.

It should be clear by now that two contradictory uses of the self-esteem concept had arrived. One was rooted in emancipation and liberation, the other in more conservative values that supported the capitalist ethos of individual endeavor and development. In the final two chapters, we will explore how the tensions between these two uses of therapeutic culture ultimately undermined the potential for collective improvement, first within the feminist movement, and then in a state-supported Californian initiative that aimed to improve the self-esteem of the entire state as a means of resolving an array of societal problems.

SEVEN

Girls and Women

In 2002, social philosopher Michael Gurian stated: "To some extent, the self-esteem dialogue has been a limiting one rather than a liberating one and it has lacked, in certain crucial areas, simple common sense."[1] Gurian was writing specifically about tendencies to caricature teenage girls as underconfident and emotionally fraught. I find this idea of self-esteem as "limiting" particularly interesting given that, in the 1960s, activists and humanistic psychologists had presented self-esteem accumulation as a boundless well of opportunity. Working on one's emotional world was supposed to be liberating, opening up possibilities for collective sociopolitical change, but the personal and political became separated. Personal transformation began to be pursued for one's own benefit, and less for any greater sociopolitical good. Instructing people that self-esteem levels were both the cause and the solution to life's difficulties encouraged people to adjust to, and accept, America's inherent inequities. In my view, this was one way in which self-esteem concept wound up being more "limiting" than "liberating."

We have seen already that the self-esteem concept appealed to multiple political agendas. Here, I explore the tensions created by this situation for feminists who struggled to agree on self-help's potential for collective improvement against the rising backdrop of consumerism, capitalism, and neoliberalism in the 1980s and 1990s. I argue that psychology could be simultaneously liberating and oppressive, and used for multiple, sometimes contradictory, purposes. Therapeutic culture was fundamentally multidimensional.[2] From the 1980s, feminist activists explored women's self-esteem issues. Most famously, journalist and activist Gloria Steinem saw empowering possibilities in the self-esteem concept for American women. Like so much of life in the 1980s, debates on beauty, fashion, and plastic surgery were recast in relation to emotions, psychologies, and, specifically, self-esteem. Feminists also considered vulnerable groups, such as schoolgirls. However, while acknowledging the existence of harmful socioemotional environments, feminists ended up wedding themselves to strategies that were fundamentally individualistic, even reactionary, creating confusion about whether feminism itself was now emancipatory or neoliberal. Nor did they fully address, or even seem to understand, the intersectional aspects of their campaigns or how different psychological life could be for groups such as African American or lesbian women. (I refrain from exploring men and boys in much detail, partly because the number of books and research published on their self-esteem at the time was relatively negligible compared to that on the female sex, and was less embroiled in activism.)

Feminist self-esteem

It's surprising that feminists barely engaged with issues of self-esteem in the 1960s and 1970s, given how effectively Black, gay, and lesbian activists had used the concept as a transformative, empowering tool. Concentrating more on their physical wellbeing, feminists challenged male medical influence in women's reproductive health and encouraged women to better understand their own bodies and reproductive systems. Self-performed vaginal, vulva, and clitoral inspections, sometimes made in public, symbolically wrested authority over female bodies from a traditionally patriarchal medical profession. A key aim of feminist literature of the time was to help women regain confidence in themselves, partly through getting to know their own bodies and being able to identify when something was wrong, an approach that got around the problem of being potentially denigrated and dismissed by male doctors. Only in the 1980s, and partly due to the influence of Black female activists, did feminist parameters of self-help expand to more fully incorporate emotional wellbeing.[3] Indeed, many leading feminists became infatuated with psychological concepts, especially self-esteem, despite potentially finding themselves enmeshed with reactionary, conservative political strategies.

Feminists had long suspected that patriarchal power damaged female self-confidence. When feminism re-emerged in the late 1960s, women sensed that collective campaigning instilled feelings of self-esteem that they did not experience in their day-to-day lives. As Peggy White and Starr Goode wrote in 1969: "New for many of us, these feelings are what gives one the courage to speak out, participate and not be afraid to fully express ourselves."[4] This new-found self-assurance was thought to support radical activity, but it wasn't particularly central. Improved self-esteem encouraged women to express their opinions and stand up for themselves, but it was not a major personal or collective end-goal in itself. Not everyone was blind to self-esteem's political potential for the women's movement. In 1969, self-esteem guru Nathaniel Branden dedicated a book to women's self-confidence which he considered to be indispensable in their fight for equality and advancement in a male-dominated world.[5] However, feminists themselves were much quieter on the subject. Perhaps they presumed that women's self-confidence would rise naturally as personal and sociopolitical prospects improved. Certainly, this was the view of female psychologist Ruth Wylie, a specialist in self-concept, who, in 1979, envisioned that the feminist movement's promotion of sex equality, female pride, and solidarity had undoubtedly boosted women's self-esteem.[6]

Another explanation rests with feminism's uneasy relationship with the psy-sciences (and, it should be said, organized medicine in general). Radical feminists spoke out against the psychiatric mistreatment of women, the power relations embedded within doctor–patient encounters, and the rampant over-prescription of tranquilizers among women.[7] As readers have seen, psychologists were indeed keen to stress the importance of motherhood, so it's little surprise that feminists attacked them for being patriarchal, conservative, and oppressive.[8] Betty Friedan and Kate Millett accused psychologists of sustaining ideologies of fulfillment which demanded that women seek emotional satisfaction in domesticity. Nonetheless, this hostility didn't stop feminists from borrowing medical and psychological terminology to expose the psychological nightmare of the mental tyranny with which they thought women lived. Even Friedan was quite partial to Rogers's and Maslow's ideas on self-actualization and its liberating possibilities for women.[9]

Recently, scholars have opted to use the term "feminisms" to encompass the multiple, sometimes competing, types of feminist activity. From the 1970s, radical feminism was eclipsed by cultural and liberal feminism before taking a turn toward individualism and the self.[10] By the 1980s, many of feminism's core aims had been met, at least partially. Growing numbers of women had entered higher education and the professional workforce, although many feminists believed that female self-esteem remained precariously low, despite these gains. Their agendas moved toward exploring the inner psychological worlds of women who had escaped domesticity's tyranny and now needed to succeed professionally. The "imposter syndrome" was one new way of understanding the lack of confidence among successful women. A well-regarded 1978 article suggested that many women with outstanding academic and professional accomplishments didn't believe they were intelligent but had nonetheless somehow fooled others into thinking otherwise. Symptoms included anxiety, low self-confidence, depression, and frustration.[11]

In 1995, Carol Eagle and Carol Colman lamented that "many feminists hoped that the women's movement, by expanding the roles and aspirations of women, would give adolescent girls a sense of empowerment that would enhance their self-esteem. But that has not happened."[12] Plunging intergenerational self-esteem was also addressed in Mary Pipher's hugely popular 1994 book, *Reviving Ophelia: Saving the Selves of Adolescent Girls*. As a clinical psychologist, Pipher found herself encountering numerous teenage girls with serious, sometimes life-threatening problems: anorexia, self-harm, suicidal thoughts. She wondered whether something had perhaps changed for the worse in girls' inner worlds in recent decades and why it was that so many girls were in therapy. She asked why so many harm themselves, pierce their faces, catch STDs, turn to drugs and alcohol. Looking at her own daughter, Pipher noticed serious mood swings and emotional turbulence. She perceived a well-adjusted girl mutating into a sad and angry failure during adolescence. Rather than behaving assertively and confidently, she seemed deeply insecure. Pipher was especially concerned about the existence of a sexualized, media-saturated culture that pressured women to appear and feel beautiful and sophisticated, a manifestation of "girl-poisoning" tendencies in modern American culture. It seemed to Pipher that, when trying to detect the root cause of their misery, most girls looked inward, wrongly considering the root cause of their problems as resting within themselves. It puzzled Pipher that female self-esteem could possibly have deteriorated to such an extent given the advances made in women's equality. Women had secured better working conditions, society was no longer as segregated. However, Pipher noticed that new sociocultural pressures had surfaced: divorced families, drugs, casual sex, violence against women – all stressors that seemed far more intense in the individualistic 1990s. Girls lost their "true self" during adolescence. Some slid into depression; others starved themselves or self-harmed; some swallowed pills, became alcohol dependent or had promiscuous sex. "Whatever the outward form of the depression, the inward form is grieving for the lost self, the authentic girl who has disappeared with adolescence. There's been a death in the family."[13]

Nonetheless, feminist perspectives on self-help were deeply contradictory. Feminists spoke out regularly about these broader sociocultural pressures, but offered solutions that discouraged attempts directly to tackle them. Moreover, Pipher's writing contained many of the key tropes and concerns of middle-class, white conservatism: the sanctity of the family, the need to guard adolescents against harmful lifestyles, unwarranted fears about rampant sexual immorality. These interests and concerns broadly mirrored those of capitalist conservatism. This was the inherent limitation of feminist liberation. Feminists acknowledged the causes of many female problems as outward but conflictingly proposed individualistic solutions to redress them. Self-help inadvertently veered feminists closer to capitalist strategies, creating frictions in feminist thought and practice.

A revolution within?

Until the 1970s, a woman's level of self-esteem was generally contemplated in relation to her domestic capabilities and familial success. Female entry into professional workplaces was restricted; society upheld domesticity as the feminine ideal. Reflecting this social milieu, psychologists examined propensities among young, underconfident women to marry unsuitable men, due to their low self-esteem.[14] Wives were expected to accumulate self-esteem through their working husbands' achievements, not their own. Women were expected to achieve and produce little, other than domestic harmony and children, but, in reality, domestic bliss was unstable. The 1950s housewife carried inside her an array of emotional pathologies. Commonly, she experienced marriage as emotionally debilitating, a source of neurosis.[15] Of course, not all marriages were unhappy, but a significant number of housewives developed mental health difficulties, sometimes stemming from a debilitated self-image.[16] Radical feminists portrayed marriage as a conspiracy against women, a male institution deliberately designed to snare and entrap them.

Nonetheless, the numbers of women entering university and professional jobs grew exponentially from the 1970s. Correspondingly, discussion of female self-image turned away from domestic life to consider how women's inner feelings aided or hindered their new public personas. In their 1984 book, *Women and Self-Esteem*, Linda Tschirhart Sanford and Mary Ellen Donovan were astonished at how little attention had been paid to female self-esteem, which they considered just as important as African American self-esteem and similarly shaped by oppression and social discrimination. For the authors, low self-esteem "constitutes an insidious form of oppression in its own right ... a woman whose career opportunities are restricted because she has been taught to think of herself as lacking in capabilities and worth is internally oppressed."[17] Sanford and Donovan thought that far too many women passed through life possessing just enough self-confidence to survive, but not enough to fully enjoy life. They reflected upon this in structural terms. Western culture deprioritized female self-esteem by deeming self-confidence to be a male prerogative while dismissing self-assured women as vain, arrogant, and conceited. Many women acted confidently in public but, deep down, harbored negative views toward themselves. Worse still, they had internalized and naturalized feelings of poor self-worth, coming to perceive low self-esteem as a natural, unalterable fact of female life, an innate aspect of feminine nature. Sanford and Donovan blamed men for encouraging women to see their self-hatred as inborn rather than psychosocial. The list of contributory factors was diverse: schools, welfare systems, domestic abuse, low pay, maltreatment at the hands of doctors, public harassment by men, empty searches for fulfillment through romance, motherhood, and work.[18] Ultimately, Sanford and Donovan depicted an oppressive world in which women were dissatisfied with their bodies, crippled by feeling inadequate, and doubtful of their talents and abilities.

Like earlier African American radical writing, feminist scholarship promoted self-confidence as a means of reorienting female identity politics. It insisted that feelings of self-worth were social (certainly not natural) and encouraged women to resist by transforming themselves from inside. Yet the feminist harnessing of self-esteem marked a curious, problematic, and contentious turning point in its sponsorship of internal, psychological solutions to external problems. Feminism seemed to be deploying therapeutic remedies popular in capitalist culture to problems that were themselves a vital component of those forces and, in doing so, were depoliticizing personal and emotional experience. This engagement with the self was criticized, even by feminists themselves, as an unwelcome retreat from radical activism. Whereas counterculture, Black, gay, and lesbian activists had adopted self-esteem-based approaches at formative stages of their campaigns, feminists, in contrast, turned to these after having already attained many of their core goals. Rather than promoting an optimistic, self-confident future for women, feminists retrospectively assessed why female self-esteem had plummeted *despite* efforts to make society inclusive and empowering for women. This was a starkly different position from that of earlier activists who took the self-esteem concept as a starting point for emancipation.

Gloria Steinem's *Revolution from Within: A Book of Self-Esteem* (1992) was the most prominent and widely debated publication that advertised feminism's inward turn. Steinem fused psychology, spiritualism, politics, self-help, science, and history. Since the late 1960s, she had been one of radical feminism's leading stars. Like Pipher, she noticed "smart, courageous and valuable" women who, beneath their confident veneer, felt far less positive about themselves. As Steinem wrote, "It was as if the female spirit were a garden that had grown beneath the shadows of barriers for so long that it kept growing in the same pattern, even after some of the barriers were gone."[19] After perusing numerous self-esteem books, Steinem concluded that these encouraged women to work on their appearance or emotions without fully contemplating the structural factors harming women's lives. She even admitted her earlier skepticism toward self-help (and psychology in general), viewing it as an oppressive force that encouraged readers to blame themselves for any unfavorable socioeconomic circumstances. She conceded that self-help placed burdens on women to constantly work on themselves: "I couldn't tell if they were protecting the status quo or just had no faith in anyone's ability to change it." Steinem remained doubtful about self-help's potential to ever tackle poverty, failure, and misery, but, regardless, she believed that self-esteem could still be fundamental to feminism. Then in her 30s, she felt as though she was neglecting an "internal center of power."[20]

The Revolution from Within presented self-esteem as a prerequisite for gender equality in democratic societies. Whereas dictatorships imposed their visions on citizens, democracies needed to nurture self-esteem in each person, creating independence, not dependence, and producing emotionally secure citizens invested in empowering others.[21] In Steinem's words:

> The problem is that if societies produce obedience by withholding core self-esteem, they are likely to discourage its mending, replenishing, and healing too. The idea of intrinsic worth is so dangerous to authoritarian systems (or to incomplete democracies in which some groups are more equal than others), that it is condemned as self-indulgent, selfish, egocentric, godless, counter-revolutionary, and any other epithet that puts the individual in the wrong.[22]

The Revolution from Within posed difficult questions about whether change should arise externally or internally, collectively or individually, and whether self-help supported only democratic forms of governance that had traditionally oppressed women. Some feminists navigated this contradiction by arguing that they had made great strides in female political empowerment, and successfully battled many outward forces, but an inner emptiness stubbornly refused to lift. Now was the time to address this.

Steinem was not alone in pursuing this path. In 1992, psychoanalyst and psychotherapist Carolynn Hillman reflected:

> My rising feminist consciousness, which started in the late sixties when I was in graduate school, certainly provided much insight into why women so often have low self-esteem. We find it hard to value ourselves because we are raised in a society that disempowers us and views us as inferior. However, this feminist analysis, while supplying some critical understanding, still left unanswered for me the question of what an individual woman could do to help herself. Yes, it would help immeasurably if society in general and men in particular valued women equally, but despite the best political efforts of many women, such enlightenment is clearly still a long way off. What are we to do in the meantime?

> As I was struggling with the issue of self-esteem in myself and in my practice, I became increasingly aware of how many of the women I was treating had a hard time nurturing themselves. They would be compassionate, accepting, respectful, and supportive towards many people in their lives, praising and encouraging them, while themselves they were berated and belittled. With experience, it became increasingly clear to me that these two issues – nurturing yourself and self-esteem were intricately linked.[23]

Hillman argued that even women with considerable educational, employment, and familial achievements still felt curiously inadequate. Western society, she argued, socialized women to attract and please others (mostly men) rather than fully value themselves. Women should "do as we're told, keep our voices down, willingly help out and mind our manners." Media advertisements reinforced messages about women's primary role being to look attractive and snare a decent man. Female self-worth came from others, not within. Hence, women needed to "challenge our upbringing and culture in order to learn to value ourselves for who we truly are." Women might have gained confidence in vocally expressing their rage but, inwardly, still believed internalized social messages about their inferiority. Being outwardly assertive was no longer enough; women should learn to respect themselves from within.[24]

This turn to self-help was fundamentally at odds with radical feminist agendas. If self-help, as its critics insisted, defused political activity and concealed the social contexts of oppression, would feminists end up being deterred from collective action? And would characterizing women as sufferers of chronically low self-esteem only end up reinforcing negative stereotypes about feminine nature? When borrowing from self-help strategies, feminists risked internalizing a sense of victimhood and seeking resolution, in an apolitical vacuum, through personal transformation. They might start thinking of themselves more as spiritual and psychological being, rather than as political and social beings.[25] They might absorb tendencies in Western psychology to depoliticize personal experience and to support, rather than challenge, the status quo.[26]

Steinem's reviewers pondered intently over whether this new phase was an advance or a retreat from political activism. In a review of Steinem's book in the *New York Times*, prominent feminist Deirdre English grumbled:

> Meditation may help or may be meaningful in its own right. What is disturbing is to see the empowering therapy supplant the cause. The strategic vision of social revolution here has all been replaced with a model of personal recovery.
>
> It isn't that Ms. Steinem has forgotten about sexism but, strangely, she has forgotten to get angry about it. Her voice seems calm and comforting, as if speaking to us from a shelter where women are nurturing women – painting, writing, laughing and singing – and never go out anymore.

English concluded that turning inward would side-track feminists from proactive campaigning. Retreating into the self, and sitting around meditating, was self-indulgent and futile. Worse still, Steinem, in English's opinion, was dangerously encouraging her female readers to consider themselves underconfident and disempowered.[27]

The feminist turn to the personal and emotional as a site of change veered curiously toward collaboration with capitalist strategies that strove to contain political struggle in the private sphere. The basic idea that low self-esteem caused oppression, and not vice versa, featured many of the hallmarks of therapeutic discourse: idealism, privatism, individualism. When applied to feminism, it risked thrusting responsibility for change and fulfillment onto individual women. Admittedly, Steinem tried hard to stretch individualism into a more collective, engaged direction. Ultimately, however, the translation of feminist politics into the discourse of the private, emotional, and psychological ended up looking ominously reminiscent of liberal-capitalist hegemony. The "personal is political" approach ended up unwittingly collaborating with contemporary capitalist culture.[28]

In 1991, Susan Faludi charted growing antifeminist and conservative influence in American culture from the 1970s. She thought that self-help books shattered the confidence of liberated women by reminding them of their underconfident personalities and then reaping the benefits by nursing their self-esteem. Having lost faith in their ability to make real sociopolitical change, demoralized women in the 1980s turned instead to changing themselves, perhaps their only option left. As Faludi wrote: "While the backlash therapy books may be written in feminist ink, they blot out the basic precept of feminist therapy, that both social and personal growth are important, necessary and mutually reinforcing."[29]

Sisterhood and political activism gave way in the 1980s to women and sisters, quite literally, "doing it for themselves." Liberation and equality collapsed into personal, private desires. Another critic, Susan J. Douglas, warned at the time:

> Narcissism as liberation gutted many of the underlying principles of the women's movement. Instead of group activity, we got escapist solitude. Instead of solidarity, we got female competition over men. And, most important, instead of seeing personal disappointments, frustrations, and failures as symptoms of an inequitable and patriarchal society, we saw these, just as in the 1950s, as personal failures for which we should blame ourselves.[30]

Perhaps we can also see here the emergence of Jessa Crispin's more recent criticisms of feminists, whom she believed turned away at some historical point from working collectively for the common female good and instead urged women to work for their own personal interests:

> What was once collective activism and a shared vision for how women might work and live in the world has become identity politics, a focus on individual history and achievement, and an unwillingness to share space with people with different opinions, worldviews, and histories. It has separated us out into smaller and smaller groups and our energy directed inward instead of outward.[31]

For Crispin, this emphasis on individualism led feminism to dissipate into little more than a fashionable lifestyle.[32]

Beautifying and dieting

To gain a better sense of how structural and individual approaches collided in feminism, it's useful to look more closely at some key sites of feminist intervention. Physical appearance was one important topic around which these debates unraveled. The 1968 Miss America protests sought to highlight how beauty ideals formed part of an oppressive patriarchal culture. In the decades that followed, feminists developed theoretically robust ideas on beauty, centered around the idea that the male-dominated capitalist economy manipulated female desires and anxieties in ways that supported the patriarchal control of women. Women were coerced into becoming obsessed with their appearance and, in doing so, they became objectified by the male gaze. Constant beautifying diverted women from intellectual work and meaningful political participation.[33]

Books such as Gerald R. Adams and Sharyn M. Crossman's *Physical Attractiveness: A Cultural Imperative* (1978) confirmed that, like it or not, physical attractiveness was now crucially important to personal development and success. For better or for worse, late twentieth-century society viewed attractiveness as a source of social influence and placed expectations, particularly on women, to attain outward perfection.[34] Feminists looked on aghast. They saw deep-rooted issues with a Western culture that counseled women to focus on physical attractiveness to attain happiness, personal success, and self-fulfillment. In 1974, psychotherapist Ella Lasky pitied "these beautiful people" who eventually found, as they aged, that "they need more than prettiness or handsomeness or their sexuality or flirtatiousness to achieve a desired goal." If beauty faded, or goals remained out of reach, such people were left "feeling empty and inept, lonely and helpless, and often very depressed." Lasky believed that culturally defined beauty standards burdened women who were forced to seek self-confidence not from work or achievement, but only from their appearance.[35] As most women ultimately failed to live up to beauty ideals, this constant cultural pressure perpetuated feelings of self-doubt.

In 1976, the *New York Times* reported on a number of women who considered their appearance unsatisfactory. One interviewee recalled: "As a homely teenager, I thought being pretty was the answer to everything. I'm thirty now, but I'm still intimidated by beautiful people." Another woman with a prominent nose, who always tried to avoid others seeing her side profile, admitted that insecurities about physical appearance had encouraged her sexual promiscuity. "I thought the only way to prove men were attracted to me was the ultimate way," she confessed; "when I have to compete against glamorous women I still feel like crawling into a corner."[36] Psychology research similarly linked physical appearance to self-esteem. One 1984 study concluded that women who wore makeup enjoyed greater confidence as people responded much better to a well-groomed appearance. Makeup might even be a useful therapeutic tool for women exhibiting "problem behavior," so this study claimed.[37] Such misogynistic perspectives seemed to confirm feminist opinions on the oppressive nature of beauty ideals.

In reality, women's emotional wellbeing usually failed to improve for long after they changed their appearance. The fragility of using physical appearance as a basis for authentic self-esteem seemed apparent when stereotypically beautiful females publicly confessed their desperately low self-esteem. Stacey Stetler was a blonde, blue-eyed, tall New York model who regularly strutted down catwalks and appeared on the front covers of fashion magazines. In 1992 she admitted that, when looking in the mirror, she saw imperfections and flaws: "I was flat-chested. You couldn't tell if I was coming or going. My back protruded almost as much as my front." Stetler opted to have breast implants.[38]

Dieting was central to the pursuit of beauty and was equally gendered. The "fight against fat" intensified from the 1950s.[39] Some feminist authors suspected that pressures to slim deepened at the very same time that women secured economic, social, and political success in the 1970s and 1980s. Doctors "medicalized" weight gain, turning it into a pathological condition previously considered natural.[40] Psychologist Rita Freedman was convinced that dieters suffered sustained psychological distress that caused more damage than their weight issues. Roberta Pollock lamented the insidious pressure to diet that created self-loathing among far too many women. She thought that body image distorted women's priorities, as weight became their central concern.[41] Meanwhile, Naomi Wolf famously insisted that, "as women released themselves from the feminine mystique of domesticity, the beauty myth took over its lost ground, expanding as it wanted to carry on its work of social control." For Wolf, beauty was the last remaining feminine ideology that still held enough power to control women, a deliberate counterforce to female advancement and part of a violent backlash against feminism.[42] The fight against fat undoubtedly also had implications for men, such as pressures to be muscular. However, the onus to be thin was placed noticeably strongly on women, and in ways that were inherently gendered.[43]

In 1978, Susie Orbach was among the first feminists to argue that women experienced the thinness ideal as "unreal, frightening and unattainable." The (then) common belief that anorexia primarily affected women led Orbach to uphold the condition as decisive evidence of the cultural pressures placed on women. If eating disorders were almost exclusively female, as Orbach believed, it seemed plausible that refusing to eat was a psychological response to women's social position.[44] In the 1980s, anorexia nervosa was a mounting problem. Critics blamed the postwar intensification of the imperative to diet, the ever-slimmer figure of the idealized female body and a renewed interest in physical fitness and athleticism.[45] For Orbach, the constant scrutiny of female bodies and the fetishization of thin "girls" persuaded women to obsess over their eating. Anorexia was a dramatic consequence, a way for women to rebel and take control of her body. In earlier times, when food had been scarcer, anorexia would have made little sense. Food was abundant in postwar America, but society encouraged women to deprive themselves of it. Orbach promoted self-help as the route to recovery to break through the "robotic response of self-hatred."[46]

In 1995, Pipher published a book on thinness, in which she claimed that "most of my female friends, both traditional and feminist, share a common disdain for their bodies" and lamented a "culture that promotes a monolithic, relentless ideal of beauty that is quite literally just short of starvation for most women." Pipher stated "to be a woman is to have a body image problem." As an alternative, she wanted positive attributes such as industry, integrity, talent, and intelligence to be used to evaluate female worth. She urged her female readers to tackle this misogynistic culture by encouraging the media to transmit different messages about women, combating harmful advertising, training doctors on eating disorders, and educating the public on the dangers of excessive dieting.[47]

However, despite this emphasis on external pressures, practical feminist advice overwhelmingly focused on personal change, through bodily acceptance, self-help, and positive thinking – a somewhat passive agenda for tackling America's beauty industries. For instance, in 1992 Mary Ann Mayo lambasted Western preoccupations with beauty for creating widespread self-doubt, and questioned whether self-esteem actually did increase all that much as women beautified. She queried the value of women seeking external cures for inner problems such as emotional wellbeing, for this seemed to end up creating more anxiety than contentment. Looking back romantically, and far too idealistically, to an idealized past, Mayo stressed that in older societies, people found contentment and inner peace by *accepting* themselves and being connected with others. In contrast, modern self-help reinforced pitiful feelings of inadequacy. This route to happiness produced only the despair of failure. Rather than boasting a positive self-image, Westerners remained burdened by negative self-evaluations. Mayo didn't want to dismantle beauty cultures but instead promoted self-acceptance, but even this remained focused on inner change rather than collective political action.[48]

Similarly, in *Bodylove*, Freedman warned that beauty ideals harmed self-esteem, and obstructed the path to female liberation. Like Mayo, she thought that even women who became optimally beautiful probably wouldn't feel much happier. Freedman's practical guide instructed that self-confidence grew when women learnt to tolerate bodily flaws and love their bodies, regardless of shape, size and appearance. Also like Mayo, Freedman declared she was not "anti-beauty"; she saw no harm in women wanting to look their best, if that's what they wanted. *Bodylove* was not about throwing away makeup and neglecting personal appearance. Indeed, Freedman saw good looks as a genuine source of women's social power. However, beauty needed to be sought joyfully, not desperately, and shouldn't be a basis for self-confidence. Playing with body image ought to be fun, not a contest. Freedman believed that all women, beautiful or otherwise, could feel satisfied or dissatisfied with their looks. Only those who learnt to view their body positively, whatever it looked like, could accrue self-esteem. Body image was independent of physical characteristics. Here again, intervention was first and foremost through positive thinking and changing one's psychological outlook. Freedman ultimately portrayed body obsession as a "contagious disease," with "bodylove" as the healing antidote, "an elixir of positive emotions, attitudes and actions." To love one's body meant to enjoy it, to view the body with pleasure. Self-esteem, not beauty, built "bodylove." The simple message was to challenge negative thoughts and stop comparing oneself to unobtainable ideals.[49] Again, cures were sought internally.

The rising popularity of plastic surgery also concerned feminists. Some practitioners actively promoted reconstructive procedures for women as "surgery for the psyche" due to its potential to boost self-esteem. Although first developed in the nineteenth century, plastic surgery did not really take off in the West until the 1970s and 1980s. Such surgery was typically elective and usually not necessary. For the most part, happiness through physical improvement was the central goal of aesthetic surgery, but many more women than men decided to go under the knife.[50]

When viewed in tandem, beautifying, dieting, and cosmetic surgery could be interpreted as women risking their health and wellbeing to find approval and feel loved. That was the opinion of Jane Wegscheider and Esther Rome who in 1996 published *Sacrificing Ourselves for Love*. They insisted that women's caring nature and centuries of subordination combined with cultural preferences for certain types of female appearances had produced an eager need to please others, specifically men. The authors thought that positive female attributes had been harmed during the historical processes of subordinating women. Their book contained various self-esteem exercises intended to empower women and safeguard their emotional wellbeing against this incessant need to please. Influence was drawn from the Fat Liberation Movement, which actively rebelled against thinness ideals. The authors viewed self-help as useful for encouraging acceptance of the body, whatever its appearance and shape, and directing enmity toward a broader culture that encouraged women to "hate and mistreat their bodies." The key message, once again, was: "Accept yourself." Rather than seek out cosmetic surgery, readers were urged to exercise their imaginations and change their self-perceptions. As body image arose in the mind, women should "teach our minds to be our allies rather than our enemies." One practical solution involved writing down negative thoughts about appearance as they arose and then challenging them with a countering positive thought. Even severer problems such as domestic abuse could seemingly be cured if readers trained themselves to believe they didn't deserve it.[51]

In the 1990s, the "born to please" idea was much discussed. Journalists described celebrities such as Tina Turner, an outwardly successful women, as suffering from the "syndrome." Turner's low self-esteem was described as stemming from her physically and psychologically abusive relationship with Ike, her ex-husband. Similarly, journalists commented on how Priscilla Presley had moved into her father Elvis's castle at the age of 14 and centered her life around him rather than develop her own independent personality. Books such as Karen Blaker's popular *Born to Please* (1990) warned women to stop being so eager to please and free themselves from controlling men.[52]

Combined, these inward-facing approaches had their limitations. A number of feminists presented women as utterly subjected to an externally imposed patriarchal culture, but others believed that the reality was less clear-cut. In 1986, Wendy Chapkis announced that women were oppressed by a "global culture machine" composed of the advertising, media, and cosmetic industries, which promoted a narrow, Westernized, and deeply oppressive beauty ideal to women all over the world. Similarly, in 1991, Faludi accused the beauty industry of destroying female bodies and minds. But, presumably, women sought beauty and slenderness for a plethora of reasons, with succumbing to overwhelming external pressures being only one of these? Moreover, it was often women themselves who helped idealize particular images of optimal beauty and perfection.[53]

In fact, women helped drive the growth of cosmetic surgery throughout the twentieth-century rather than passively being the subjects of its coercion; they were eager consumers, not hapless victims. Kathy Peiss argues that the beauty business was largely built *by* women, rather than solely *for* them. Just as consumers could be both active and passive, beauty culture could be simultaneously coercive and liberating, presenting yet another paradox. As Kathy Davis, herself a feminist, added at the time, the appeal of cosmetic surgery was more complicated than feminists suggested. Rather than being brainwashed victims of the media, women were active agents who negotiated their bodies and lives within the cultural and structural constraints of a gendered social order. Many viewed cosmetic surgery as a positive personal choice that allowed them to positively take control of their lives.[54]

Of course, it's possible to argue in response that such women are being duped and coerced without realizing it. Nonetheless, English professor Virginia Blum, in her 2003 study of the culture of cosmetic surgery, contended that feminist criticisms of bodily practices usually ended up being just as shaming as the physical imperfections that drove people to beautify in the first place. For Blum, too many feminists hold themselves up as superior to the cultural machinery, while watching others "desperately fling themselves across the tracks of cultural desire." She insisted that some feminists went overboard in berating women's concern with outward appearance, largely overlooking the pleasures achieved through paying positive attention to bodily appearance. Women did things for pleasure, and not always out of shame. Moreover, feminist attacks on the beauty industry failed to convince a majority of women to stop buying cosmetics and fashion magazines, for beauty had positive purposes for self-definition.[55]

Here we can see how the pursuit of self-esteem simultaneously contained both limiting and liberating possibilities. This contradiction was visible in Elizabeth Taylor's autobiography *Elizabeth Takes Off*, published in 1988, in which she reflected upon her well-publicized weight gain during the 1970s and 1980s. She recalled that her emotional life plummeted as she approached her forties, along with her physical appearance. Overeating became a way for her to compensate for her flailing self-esteem, but the subsequent weight gain caused her self-confidence to disintegrate still further. As she wrote: "We all know heavy people whose emotional lives are in good order who have an excellent sense of self-worth. And Hollywood is filled with thin, beautiful women who are unhappy and unfulfilled, with little sense of self-esteem." Taylor made a distinction between how she looked and how she felt. She also urged her readers to take charge of their self-image and weight loss. For her, self-motivation was the only answer: "Unless you can turn the switch on in your own mind, you won't succeed." Self-destructive habits needed to be thrown away and a new way of life embraced.[56]

Taylor's autobiography can be interpreted in various ways. On the one hand, it could be read as striking evidence of the sociocultural pressures placed on women to be slim, which Taylor had unwittingly succumbed to. But, then again, her experiences of dieting and restoring her figure could be seen as her being experienced and feeling positive. There were few signs of feeling oppressed or unwillingly victimized by beauty standards. For Taylor, weight loss, rather than self-acceptance, was empowering and self-esteem boosting.

Of course, it could be said that the feminist turn toward the self failed to recognize that anorexia, dieting, and excessive plastic surgery were solely patriarchal issues. Beauty standards, beyond doubt, helped oppress women to some extent, yet a broader problem has existed that forms the very heart of this book: the Western obsession with what others think of us and our physical appearance. The true source of oppression might well have been a social system that requires us all, male or female, white or Black, to constantly pay attention to what others think of us. Perhaps it is possible to look slightly beyond feminist debates to detect a more overarching, fundamental problem in our modern culture: selves that depend so intimately on the opinions of others are fragile.[57] A fundamental question in relation to beauty-related matters is not whether it is feminist or antifeminist but why it has been that so many women became convinced that their physical features were so undesirable that they sought help.[58] If the answers can truly be found in the broader emotional environments inculcated by capitalism, then it becomes clearer why individualistic approaches have proven so thorny for feminist strategies.

School girls

Education was an ideological battleground in which feminist debates about self-esteem were played out and it provides another illuminating case study. Additionally, it exposes problems in feminist portrayals of underconfident females and engagement with intricate debates on gender difference. In its widely discussed 1991 report, *Shortchanging Girls, Shortchanging America*, the American Association of University Women (AAUW) claimed to have "awakened the nation to the effects of gender bias in American schools." Its report was underpinned with interviews conducted with girls and boys aged 9 to 15, with an emphasis on self-confidence, academic interests, and career goals. Between September and November 1990, interviews were conducted with 2,374 girls and 600 boys. The report concluded that both sexes suffered a decline in self-esteem at the start of adolescence, but for girls this decline was more dramatic and longer lasting. Eight-year-old girls were usually confident and assertive, but during adolescence a majority of them began suffering negative self-image and diminished hopes for the future. As they became mindful of their limited future career options, self-esteem dwindled. The self-esteem of boys, meanwhile, envisioning glowing future careers, grew exponentially.[59]

The AAUW argued that schools reinforced gender-biased social messages. Despite entering first grade with skills and ambitions similar to those of boys, girls ultimately left high school with disproportionately low self-confidence. The AAUW accused American schools of "cheating" and "short-changing" girls by stressing competitive rather than cooperative learning and using textbooks devoid of female role models, all of which reinforced negative stereotypes about female ability. Not that this was a deliberate conspiracy against girls. Teachers *unconsciously* dampened girls' aspirations, particularly in math and science, ultimately lowering women's contribution to American society. This allusion to core math and science skills injected economic and political urgency into the issue: America, the AAUW warned, risked losing its globally competitive edge. With ever-growing numbers of women entering the workplace, what would be the impacts of having half of America's workforce unsuitably trained?[60] Of course, this discussion of girls' careers only arose because women's employment opportunities had recently improved. Feminists now spent less time fighting to enter professional workplaces than to exposing the new barriers that faced them.

The AAUW report attracted major international interest and journalistic exposure. Washington professor and feminist Myra Sadker reported to the *New York Times*: "It's really quite staggering to see that this is still going on. No one has taken such a large-scale look at self-esteem before but we have known of this issue for years. And here you see that it is not going away."[61] While the AAUW had presented educational sexism as mostly unconscious, Sadker, with David Sadker, pronounced a "curriculum of sexist school lessons becoming secret mind games played against female children, our daughters, tomorrow's women."[62]

Carol Gilligan also engaged with the debate. Gilligan is an influential feminist who spoke out against women sacrificing their lives and wellbeing to become good wives. She is well-known for her 1982 book *In a Different Voice*, which explored female psychology and girls' development.[63] In 1990, she responded to the AAUW report:

> This survey makes it impossible to say what happens to girls is simply a matter of hormones. If that was it, then the loss of self-esteem would happen to all girls and at roughly the same time. This work raises all kinds of issues about cultural contributions and it raises questions about the role of the schools, both in the drop of self-esteem and in the potential for intervention.[64]

Gilligan considered the transition from girlhood to adolescence to be a critical juncture in which girls witnessed a "relational crisis in women's psychology." She presented girls in general as victims of incest and sexual abuse, susceptible to depression and trauma. She contributed to perceptions of adolescence as a time of turmoil in which girls became disconnected from their true self. Gilligan considered this experience to be so traumatic that most women couldn't remember it happening to them. They blocked out the trauma.[65]

The AAUW report inspired a number of authors, some allied to the Association, to pen provocative books. In 1990, Berkeley psychologist Emily Hancock published *The Girl Within*. Based largely on personal interviews with youngsters, she argued that girls' self-esteem peaked around age 9, then rapidly plummeted. Girls lost their prepubescent strength and independence to become overly preoccupied with personal appearance. At this transitional life stage, they began comparing themselves to waiflike women, ones with tiny waistlines – a first step, Hancock warned, toward eating disorders, depression, suicidal thoughts, sadness, smoking, and alcohol and drug abuse.[66] Hancock, a self-confessed disciple of Gilligan, considered pre-adolescence as a period of least exposure to patriarchal forces. Rediscovering the "lost" pre-adolescent girl would restore and reinvigorate positive female adult identity. This approach resonated with Steinem's and Pipher's ideas on the need to recover identities lost during girlhood while being overwhelmed by men, the patriarchy, and prescriptive patterns of womanhood. Women's true self needed to be refound; a feminine world thought to have survived separately, and internally from the patriarchy's devaluation.[67]

In *Schoolgirls*, Peggy Orenstein also expanded upon the AAUW's report, depicting a gloomy world in which "teenage girls are more vulnerable to feelings of depression and hopelessness and are four times more likely to commit suicide." Orenstein portrayed teenage girls' lives as revolving almost entirely around self-image, self-doubt, and self-censorship. Like Steinem, she admitted to having once avoided the self-esteem movement because it favored personal transformation over structural change. She had also once regarded self-help as victim-blaming, a means for men to sell women's self-esteem back after having stolen it in the first place. (Of course, this conceals the fact that many women wrote self-esteem manuals.) However, Orenstein's practical work with young girls convinced her that the inner world should not be ignored after all. Like Pipher and Steinem, Orenstein wondered how female self-esteem could have fallen so low in a more gender equitable America. Women still felt unsure, insecure, inadequate. Indeed, writing *Schoolgirls* helped Orenstein confront her own conflicts. She came to realize, after observing others do the same, that she had been suppressing herself, censoring her own ideas and doubting the efficacy of her own actions.[68]

Despite its influence, the AAUW report had many critics. Instead of using a well-established self-esteem measuring scale, the group designed its own, leaving its quantitative basis open to dispute. The pre-adolescent sample size was smaller than the norm in studies of young people's self-esteem. Data had been collected at a single point in time, offering little evidence of how self-esteem changed across the adolescent life cycle.[69] Even if the AAUW had used standard measurement scales, it might not have found the patterns hoped for. Conclusive evidence of significant difference between male and female self-esteem remained persistently elusive.[70] Thorough overviews of self-esteem such as Ruth Wylie's 1979 two-volume book *The Self-Concept* had criticized the "conceptual carelessness" common in studies of gender and self-esteem.[71] Just three years before the AAUW published its report, researchers still commented on the sparsity of decisive empirical evidence on gender difference in self-esteem levels. Whether or not self-esteem dropped due to changes in school environment or psychobiological changes remained unclear. The report also failed to reflect in any detail on whether changes in self-esteem varied much in people of the same gender. Had too many broad conclusions and generalizations been reached? (Even in the twenty-first century, academic research still remains inconclusive on the matter of self-esteem and gender difference.)[72]

The international publicity garnered by the report encouraged close scrutiny of the evidence, as well as the ideas on female self-esteem being discussed more broadly by Pipher, Steinem, and others. Some critics queried these authors' underlying political motivations and how they might be subjectively shaping the research findings. In 1990, educationalists James Kenway and Sue Willis counterargued that "the claims being made on its [self-esteem] behalf as an explanation of school failure and as an elixir for both school success and the liberation of girls seem simplistic to say the least." Kenway and Willis framed self-esteem as a "regime of truth," one that some critics felt afraid to speak out against in case they appeared opposed to happiness itself. They criticized the "scanty research evidence" underpinning feminist school-based programs and tendencies to consider girls' self-esteem universally, ignoring their specific cultural circumstances of girls, diversity, and the lack of attention paid to factors including race and class.[73]

Feminists themselves were among the most vocal critics. In 1994, Christina Hoff Sommers accused the AAUW of "aggressively promoting the sensational but empirically unfounded thesis that gender bias is causing a debilitating loss of self-esteem in our nation's schoolgirls." Sommers described herself as an "equity feminism," not (what she termed) a "victim feminist" or a "gender feminist." She perceived an irrational hostility toward men in many strands of feminism and an inability to grapple constructively with the possibility of biological and psychological difference. Sommers took issue both with the anger of many feminists, but also with Steinem's calmer descent into personal introspection. Unsurprisingly, she publicly questioned the AAUW's motivations and claims to have discovered "an unacknowledged American tragedy." She lampooned the group's use of T-shirts and mugs to promote its cause and contested arguments in Orenstein's *Schoolgirls*. Cited in the *Wall Street Journal*, Sommers furiously wrote:

> Consider this major piece of evidence adduced by the AAUW to highlight the difference in boys' and girls' aspirations for success: "The higher self-esteem of young men translates into bigger career dreams ... the number of boys who aspire to glamourous occupations (rock star, sports star) is greater than that of young women at every stage of adolescence, creating a kind of glamour gap." A glamour gap? Most kids do not have the talent and drive to be rock stars. The sensible ones know it. (The number one career aspiration among girls, by the way, was lawyer). What the response of the children suggest, and what many experts on adolescent development will tell you, is that girls mature earlier than boys, who at this age, apparently, suffer from a "reality gap."

To support her critique, Sommers cited official government statistics which suggested that, in reality, many more boys than girls suffered from learning disabilities, delinquency, alcoholism, drug abuse, and suicidal tendencies, as well as lower grades. She concluded:

> In sum, a politicized AAUW has effectively used its own advocacy research to promote the myth of a pervasive demoralizing bias against schoolgirls. It

has created a cult of the persecuted girl-child among feminist writers. And it has gotten legislators to allocate millions to address a "gender gap" in the nation's schools.[74]

Sommers was particularly concerned about inaccuracies. Guided by clear political agendas, some feminists undoubtedly exaggerated, oversimplified, and obfuscated. However, Sommers saw this lack of rigor and scholarly care as undermining feminism, leaving it vulnerable to easy criticism and pot shots. She likened some feminist groups, including the AAUW, to untrustworthy advocacy groups – the tobacco industry and National Rifle Association, for instance – which similarly presented deeply biased information to the public. Sommers doubted whether most women, or even many feminists, viewed contemporary America as quite the oppressive space of overwhelming patriarchal oppression portrayed by many feminists. Ultimately, she dismissed the AAUW as a "para-academic organization that has become highly political and ideological in recent years," and blamed the current leadership for turning the group into an "activist arm of gender feminism." She accused the AAUW of ignoring, perhaps deliberately, a substantial body of research that failed to reach decisive conclusions on gender and self-esteem, instead relying heavily on its own scholar, Gilligan, whom Sommers considered to have less standing among her academic colleagues than among feminists.[75]

The strong assertions about female adolescents' self-esteem quickly unraveled. The AAUW attempted to tackle a broader sociocultural and environmental factor: the education system. Nonetheless, the close scrutiny awarded to its methodologies and opinions undermined their claims. The AAUW campaigned for useful practical changes, ones that were presumably more understandable and accessible than abstract psychoanalytic ideas of relocating lost girls. But fundamental problems in the empirical evidence, combined with tendencies to caricature and universalize girls' inner lives, weakened the project.

Biological or environmental difference?

Twenty-first-century authors on gender difference later criticized tendencies of the 1990s to portray young American girls as helpless targets of discrimination, cheated out of their education. Strong evidence has not been forthcoming. Indeed, most research concluded that girls' school performance excelled that of boys.[76] Looking back retrospectively, psychologist Rosalind Barnett and journalism professor Caryl Rivers worried in 2004 that "parents may have gazed at these images with horror while underestimating the resilience of their daughters, who were able to thrive in spite of them." They also dismissed the "crisis" as a myth that caused undue anxiety in parents and popularized misguided ideas such as single-sex classrooms. Additionally, they queried Pipher's overreliance on case studies rather than systematic quantitative research.[77] A new breed of gender researchers re-examined the authors cited above and bemoaned the pitiful constructions of female weakness and vulnerability and willful glossing over of low self-esteem among teenage boys.[78]

A rigid dichotomy had been drawn between male and female pupils that established gender difference as being socially and culturally, certainly not biologically, determined. However, this risked veering into a strict environmental determinism. Many feminists refuted suggestions that men and women might, even in some ways, be biologically or psychologically different. These issues fed into a broader, deeply contested debate about gender/sex difference. Indeed, the issue of whether the sexes are fundamentally different, physiologically or psychologically, remains heatedly debated. Traditionally, feminists maintain that gender differences are primarily social and that male/female brains are more or less the same. Cognitive psychologist Steven Pinker has argued, causing much debate, that after the Holocaust, a tragic event driven largely by biological determinism, it became unfashionable to suggest that human behavior might be biologically driven. To claim so risked returning to earlier times when scientists had rooted female (and racial) inferiority in the distinctiveness of biology and physiology. In the postwar period, environmental determinism took prominence and framed human behavior in relation to sociocultural environments. The idea that biology might strongly influence day-to-day behavior became tantamount to blasphemy.[79]

Many believe that feminists went too far in rejecting nature and, more specifically, biological and psychological differences between the two sexes.[80] Another line of thought professes that the two sexes are indeed different and that recognizing this might have benefits. Others contend that generalizations cannot be made about either women or men as a group, as both sexes demonstrate considerable variation from person to person and in different situations.[81] The inherent complexities of these debates indicate that feminists interested in girls' self-esteem subscribed to approaches that refused to entertain suggestions that female education was anything but sociocultural. On a practical level, the gloomy depictions of adolescent girls' emotional wellbeing was at odds with the lived realities of girls. Caricatured images of sad girls and happy boys ignored the sheer diversity of youthful experience. Scholars were later horrified at the messages distributed to adolescent girls and their parents in the 1990s.

But, in many ways, this tendency to pathologize, victimize, and generalize about those with low self-esteem can be read as yet another manifestation of the processes by which those lacking self-confidence were encouraged to blame themselves and assume responsibility by mastering their feelings, part and parcel of conservative strands of therapeutic culture. In 1994, Myra and David Sadker depicted 6-year-old girls as full of energy, self-reliance, purpose, and confidence, all of which was destroyed, "like the tightening of a corset," by adolescence. The transition from elementary to middle school was identified as a time of trauma, one that contributed greatly to the shattering of girls' fragile self-esteem. Almost all teachers reinforced this process by ignoring or dismissing girls, "short-circuiting" the psychological processes by which healthy self-esteem was acquired. Adolescent girls then placed them at emotional and academic risk. As the Sadkers wrote: "No longer is she at the peak of her world, in charge of all she surveys. Instead she walks cautiously, wary of the traps around her. She moves on to high school with all its promise, but she begins the journey on very shaky ground."[82]

In such startling proclamations, female experience became almost caricatured, the lives of girls reduced to misery and lack of confidence, their perfunctory education forcing them to grow into women unable to like themselves. However, the magnitude of gender difference in self-esteem was probably less consequential than feminists thought. Indeed, in the twenty-first century some skeptics insisted that single-sex schools create as many problems as they resolve and that most girls do not require special protection or self-esteem workshops after all.[83] Feminist claims undoubtedly contributed to a broader political agenda, one intent on positioning gender difference as socially and culturally constructed;

but depressing images of underconfident, substance-abusing, soon-to-be pregnant teenager failed to map onto the realities of life for many teenagers and their parents.

There existed a bigger picture problem. Sociologist Dana Becker argues that "psychological woman," as she terms it, is by and large not an activist, as therapeutic culture tends not to create individuals directed toward changing the status quo. Instead, therapeutic culture offers women "a type of compensatory power that supports and reproduces the existing societal power/gender arrangements by obviating the need for social action to alter them." Since the 1970s, Becker believes, female campaigners became more attracted to ideas about the self, rather than larger political realities, with therapeutic language offering them a means to express themselves.[84] This context shaped ideas about female self-esteem in the 1980s and 1990s.

Feminist discussion of self-esteem ultimately individualized, pathologized, and depoliticized sexism, mirroring capitalist tendencies in self-help advice. Although well-meaning, it informed young girls and parents that they sorely lacked attitudes, skills, and personal attributes and needed self-esteem therapy to feel better about themselves, discouraging girls from fully considering the broader structural aspects of their adverse feelings. The imperative to self-improve meant that those who refused to do so still risked being blamed for their own poor educational outcomes. As one critical 1990 essay argued: "In proffering its somewhat deficit view, the literature rather perversely constructs its own stereotypes of girls and these look strangely similar to those which it seeks to defy."[85]

While Steinem, the AAUW, and others admittedly acknowledged the broader social contexts of low self-esteem, their advice to personally improve failed to promote a sense of sisterhood or community as second-wave feminists had earlier done. Like liberal feminism, it implied a social benefit in collective self-interest, directing individuals inward rather than outward. The starting point was always the individual, not the collective.[86] Pipher's *Reviving Ophelia* came under particular attack for propagating myths that girls' lives were dominated by distress caused predominantly by a misogynistic male-dominated society.[87] As Gurian wrote:

> For my daughter's sake, I must ask: What happens to a culture that promotes the idea that males are inherently defective, violent or woman-hating, and

females are inherently victims? How will my daughters make the compassionate alliances they need when they are adults if they are trained to believe boys and men are predominantly destructive to them?[88]

A further, perhaps predictable problem, was that feminists inadvertently idealized the self-confident girl (and woman), one who was essentially white and middle class. If Black women and girls feel ominously absent from this chapter, that's because they were also absent from my source material. As the debates about the AAUW's report raged on in the 1990s, psychotherapist Julia Boyd, writing on Black women's self-esteem, wrote:

> The main thing I learned in school was that I didn't exist. Not as a student, a female, or a person of color. What I learned was that the American educational system, like much of the society we live in, was never intended to educate or support our female gender or our ethnic diversity. I also learned that white schools were created, designed, and implemented by white males in order to educate white males ... it's really not surprising that a high number of young Black females fail and are pushed out, as opposed to drop out, of an educational system that was never designed to include us ... While there might not seem to be a connection between education and personal self-esteem, they are powerfully linked.[89]

If we reflect critically on feminism's problematic alignment with the values of therapeutic culture, it now seems less surprising that it also absorbed, knowingly or otherwise, the color-blindness rampant in white American culture, particularly since the 1970s.[90] From then on, feminist psychologists began considering how psychologies shaped women's lives; however, in doing so, they created a psychology that automatically "othered" nonwhite women. Despite overturning the white, middle-class male as the norm, feminist psychologists didn't always study women in their different contexts and socioeconomic conditions.[91] What some call "neoliberal feminism" emerged, a self-help program for upper-middle-class women that left behind those who weren't privileged professionals.[92] Attorney and feminist journalist Rafia Zakaria, in her condemnatory 2021 book, accuses "white feminism" of ignoring, erasing, or excluding Black voices not necessarily consciously, but because these women benefit from white supremacy. Zakaria's key target is "whiteness within feminism," rather than white women per se. Feminism is color-blind when it fails to observe the white privilege embedded within it. More specifically, she lambasts the types of feminism that emerged from around the 1980s for encompassing capitalism's individualistic ethos. In particular, she targets Steinem's *Revolution from Within* for reducing women's problems to poor self-image.[93]

Despite being engrossed with girls' education, little was being said about the lack of Black female figures and role models visible in the classroom. Where Black people occasionally featured, they were usually male. A clear message was transmitted: Black women didn't exist. Boyd remembered performing well at school, but the lack of Black-oriented education "just made the transference of our educational process to our personal reality more difficult." Boyd was acutely aware of the double pressures placed on her for being both Black and female. She advised Black women against defining themselves as "the wounds of the past," for "healthy self-esteem is a present reality." Like many African American writers of the time, she spoke of the need to "survive" somewhere outside mainstream society.[94]

Looking further back, bell hooks observed in 1982 that Black women were largely silent in the women's liberation movement. Many sensed that their racial identity, and its resultant discrimination, was more significant to their lives than gender issues. However, the Black Power movement had hardly been known for its glowing attitudes toward Black women, having essentially sought to install a Black patriarchy. And hooks agreed that many white feminists harbored sexist-racist attitudes toward Black women. Even those who didn't act overtly racist had probably absorbed, supported, and advocated racist ideologies. For hooks, second-wave feminism was a means through which white women voiced their grievances, while treating women of other races as "others."[95] She also believed that poor women were largely excluded from feminist narratives:

> It was not in the opportunistic interests of white and upper-class participants in the women's movement to draw attention to the plight of poor women, or the specific plight of Black women. A white woman professor who wants the public to see her as victimized and oppressed because she is denied tenure is not about to evoke images of poor women working as domestics receiving less than the minimum wage struggling to raise a family single-handed.[96]

In turn, Black women rejected feminism as it seemed too white and middle class, and too supportive of race and class hierarchies. Calling into question notions of sisterhood, hooks wrote that feminists wanted to enter mainstream capitalist society, and not so much challenge it.[97] Such themes have emerged in recent writing on feminism by Black authors such as Mikki Kendall, who argues that much feminism is laden with racist and classist assumptions about Black women. For Kendall, mainstream feminism has essentially served white, middle-class female interests and showed little interest in the basic day-to-day needs of working-class women, particularly those from Black communities.[98]

Evidently, feminist debates on the nature of female self-esteem (or lack of it) has been awash with debates, disagreements, and controversies, and the discussion continues to this day. For the purposes of this book, it seems apparent that a rift developed within feminism partly because of the underlying tensions between the structural and the personal, and because of self-esteem's uneasy ability to appeal simultaneously to both radical and conservative politics. Feminists such as Steinem truly believed that introspective self-help techniques could serve a useful purpose in women's collective liberation, but found it more or less impossible to navigate the potential of psychological self-help ideas to internalize personal problems that might well have benefited from greater attention being paid to the structural, social, and political. Chapter 8 continues to explore this dilemma by assessing what happened in the late 1980s when the state of California attempted to implement initiatives intended to boost the self-esteem of its citizens, with the ultimate goal being to tackle the state's various social problems.

EIGHT

A Self-Esteem Task Force

In 1986, the self-esteem movement received a boost with the establishment in California of the Task Force to Promote Self-Esteem and Personal and Social Responsibility. Launched to raise the self-esteem of Californians, it hoped to resolve the plethora of social problems which, as previous chapters have shown, were being blamed upon low self-esteem. Led by John Vasconcellos, the Task Force took for granted that low self-esteem preceded various behavioral and social problems (rather than vice versa), lending much urgency to the subject of Californian self-esteem levels. The Task Force can be seen as the zenith of the self-esteem movement, a historical moment when politicians allocated state funding to support practical initiatives based on a stubbornly intangible psychological concept.

However, the Task Force's work was beset by various problems in its research designs, ambitions, and debatable framing of low self-esteem as cause, not consequence, of American societal problems. Like feminists, the Task Force struggled to surmount the paradoxical trappings of offering individualistic solutions to structural problems in the new capitalist, neoliberal era. Emulating mainstream self-esteem, and entirely ignoring the work of activists on promoting self-esteem, the Task Force paid only lip-service to socially marginalized groups, reflecting therapeutic culture's tendencies to generalize about its targets and downplay the harmful impacts of social modernity and discrimination on emotional wellbeing. It was almost as if decades of African American, gay, lesbian, and feminist work on self-esteem simply hadn't happened. But despite overlooking the successful work of activists, and highlighting specific psychological needs among specific groups, the Task Force claimed to be working in the interests of all Californians. Ultimately, the international attention garnered by the Task Force encouraged a public questioning of the self-esteem movement, its aims and approaches, and its flawed outlooks.

The establishment of the Task Force is a compelling episode in self-esteem's history, one in which the ideas, practices, and vocabularies of the therapeutic enterprise were put to practical use, enabling active forms of governance centered around positive emotions. By the 1980s, the main questions being asked were whether citizens were "happy" and "healthy," not, as in earlier times, "good" or "bad," or "hungry" or "well-fed." Unlike traditional religion, the new therapeutic enterprise was self-referential. As part of their sociomoral duties, Californian citizens were urged to look within, to improve themselves.[1] But, just like the feminists, Vasconcellos struggled to reconcile the contradictions inherent in the self-esteem concept. Predictably, the Task Force ended up aligning more closely with conservative and capitalist agendas. Once again, the boundaries between the self-esteem concept – as oppressive or liberating, limiting or limitless – became blurred because the idea was so vague and flexible that it could be adopted to support inherently contradictory political agendas.

A new task force

Vasconcellos was proudly influenced by humanistic psychology.[2] Similarly rejecting psychology's earlier focus on negative psychopathologies, he became persuaded that humans could "self-actualize" and enjoy lives filled with happiness, creativity, and fulfillment. Indeed, he believe that this should be actively encouraged, for the end result would surely be a better society populated with happy, confident people. The Human Potential Movement's flagship institution was the Esalen Institute, established in 1962 in Big Sur, California.[3] Vasconcellos worked there for twenty years as a therapist before initiating the self-esteem Task Force. In the meantime, he rose in prominence in Californian politics. In 1986, the *New York Times* described him as "the most radical humanist in the Legislature."[4]

Nonetheless, times had changed. Vasconcellos was promoting his ideas in the 1980s, when self-help strategies had been co-opted to support rightwing political agendas rather than radical social change. For some, humanistic therapy still promised liberation from capitalist modernity.[5] Nonetheless, its basic principles could still contradictorily support conservative values: individualism, self-sufficiency, familial stability.

Rather than offering liberation, these limitations encouraged everyone to assume blame and responsibility for any problems that came their way. Hence, Vasconcellos faced the same dilemma as the feminists: how to reconcile the contradictions inherent in the self-esteem concept and how to navigate its emancipatory and oppressive potentialities. Vasconcellos might have learnt much from activist groups such as Black Power, the gay liberation movement, and, to some extent, the contemporary feminist movement, but instead he remained largely blind to race, sexuality, and gender, advocating a more generalized approach.

Political theorists disagreed on whether the state ought to assume responsibility for improving everyone's emotional wellbeing. Philosopher Karl Popper was doggedly opposed to totalitarianism and collectivism, and believed that it wasn't the government's business to actively make citizens happy, although reducing misery and suffering might be. Writing in the 1960s, Popper warned that official efforts to promote happiness (including self-esteem) would infringe personal liberties. In the 1970s, philosopher Robert Nozick similarly argued that self-esteem should be privately dealt with. However, political philosopher John Rawls, writing in the early 1970s, insisted that self-esteem was a public concern, and should be socially distributed in ways that protected individual rights. Rawls's model granted governments jurisdiction over the circumstances that promoted self-esteem. Inauguration of the self-esteem Task Force saw this philosophical perspective put into practice.[6]

Vasconcellos first attempted to mandate his Task Force into law in 1984, but a heart attack delayed his plans. After recovering, he tried again, although this attempt was vetoed by the state governor. Vasconcellos then renamed his group: "The Task Force to Promote Self-Esteem and Personal and Social Responsibility." The addition of "personal and social responsibility" marked a move toward promoting self-esteem as beneficial to the public interest. Practical considerations undoubtedly played a part too. Self-esteem could only successfully find a legitimate place in policymaking and secure public funding if it was presented as a cost-effective solution to social problems in the Reaganite era, which it certainly promised to be. After all, inner emotional work is cost-effective.[7] The inclusion of "responsibility" rebalanced self-esteem's individualistic orientation by sponsoring a civic republican ideal of commitment to the general good. However, the basic point of reference remained the self.[8]

Vasconcellos portrayed the Task Force's work as a "new vision to build a society in which self-esteem is nurtured and people naturally assume personal and social responsibility."[9] He presumed that empowered, confident people would naturally begin to engage in matters relating to their community's welfare, while those still burdened by low self-esteem would continue engaging in socially costly, criminal behaviors. Given this, intervening in everyone's self-esteem levels seemed all the more urgent. If self-esteem truly did lead to so many social problems, as most people had come to believe by the 1980s, then politicians surely had responsibilities to nurture self-esteem and "rehabilitate" those suffering from low self-worth?[10]

On September 23, 1986, a new Bill was passed almost unanimously (aside from opposition from the State Department of Mental Health, which considered self-esteem to fall under its own purview).[11] The Task Force was allocated a budget of $735,000. Despite only having space for twenty-five unpaid volunteers, the board received 350 applications. This was no conventional board. Its first meeting began with a female speaker discussing her book *Gourmet Parenting*, another author presenting on his self-help program entitled "How to be Successful in Less than Ten Minutes a Day," and the group's chair, anthropologist Andrew Mecca, discussing his dream-walking experiences with Aborigines. One board member predicted that the Task Force's work would be of such importance that it would cause the sun to start rising in the west.[12]

Soaring aspirations indeed. However, initial progress was slow. The board spent an entire year disagreeing about how to define self-esteem. After considerable discussion, they settled on: "Appreciating my own worth and importance and having the character to be accountable for myself and to act responsibly toward others." Then, the board needed to determine which Californians had self-esteem, who didn't, and the reasons why. Its members spoke to a broad spectrum of individuals, including street gang members, single mothers, counselors, and community leaders. The Task Force also developed research agendas to establish "whether healthy self-esteem relates to the development of personal responsibility and social problems (including crime, alcoholism and violence) and how healthy self-esteem was nurtured, harmed or reduced, and rehabilitated."[13]

The Task Force focused initially on seven key areas: families, education, drug and alcohol abuse, crime and vice, workplaces, poverty, and chronic welfare dependency. It portrayed healthy self-esteem as a "social vaccine" that would empower individuals to live responsibly and "inoculate" them against all manner of problems including teenage pregnancy and educational failure.[14] Raising self-esteem was a magical remedy, a means of creating socially desirable behaviors. Seemingly, reconfiguring everyone's self-image was a powerful enough tool for curing modern society's ills. If the project failed, blame could easily be passed on to individuals for having failed to summon up enough positive emotions. In these ways, an ostensibly emancipatory, liberal project became embroiled with the burgeoning capitalist culture of the day.[15] As we have already noted, self-esteem was at once liberating and oppressive, limitless and limiting.

An out-there idea?

California was at the heart of the counterculture, but had gained a reputation as the home of kooky, "out-there" ideas. Initially, the Task Force was internationally ridiculed. Famously, American cartoonist Garry Trudeau spent a fortnight writing satirical comic strips about it in his popular Doonesbury comic strip series. In Britain, *The Times* reported on the "weirdness of the feel-good state," adding that "the rest of the country [America] has yet to stop laughing." The *San Francisco Examiner* commented: "As if they had needed to reinforce Sacramento's credentials as the kook capital of the world. The legislature's willingness to underwrite screwball ideas like this make it difficult for a Californian to keep his self-esteem."[16]

Facing global mockery, Vasconcellos insisted that Californians were hugely interested in the scheme, claiming that, even before its first meeting, the Task Force had received more than 2,000 calls and 400 volunteer applications. The hype surrounding the scheme mushroomed. For intended and unintended reasons, it attracted huge public interest. Vasconcellos was invited to speak on television shows internationally.[17] To boost its credibility, the Task Force also recruited seven professors from the University of California, hoping to compile "the world's most credible and contemporary research regarding whether healthy self-esteem relates to the development of personal responsibility and social problems," including crime, drug abuse, teenage pregnancy, and welfare dependency.

By the 1980s, ideas about self-esteem were well embedded in school curricula. In light of the Task Force's unexpected success, schools across America added further self-esteem material. By Spring 1990, 86 percent of California's state elementary school districts and 83 percent of high school districts had implemented, or were planning, new self-esteem programs. Students in San Antonio, Texas, were graded on subject matter, as well as honesty, truthfulness, kindness, courage, and conviction. One teacher, Jack Canfield, began to run seminars for his colleagues, which included motivational exercises. Typically, these commenced with participants massaging their colleagues' shoulders to relax the group and create bonds. Then, in small groups, the participants were urged to recall successes in their lives or describe the highlights of their lives in five minutes. "It's been a remarkable shift in this movement since I first got involved in 1984," announced Vasconcellos in a 1990 *New York Times* interview. "People are exasperated," he claimed. "Nothing else has worked, so they're willing to listen to new proposals."[18] That same year, Vasconcellos added: "We have moved self-esteem from an object of Garry Trudeau's satire to a subject widely and deservedly respected."[19]

But not everyone was convinced. In 1990, Gary Bauer, president of the Family Research Council, pre-empting a rise in narcissism, insisted: "We certainly are for self-esteem. It's a good thing for kids, particularly, to have confidence in themselves. We do feel, however, that there are a lot of odd things going on there in the name of self-esteem. And in some cases, self-esteem degenerates into just another form of almost selfishness." Roger Wilkins, a professor at George Mason University, expressed a further pointed concern: "For those born into or living in situations where there is no stability, or where the only successful people are criminals, self-esteem isn't going to do it for them." He recognized that self-improvement strategies often glossed over deep-rooted and highly complex social problems that demanded more intricate solutions than persuading everyone to feel good about themselves.[20]

A social vaccine

In January 1990, the Task Force published its 161-page report, *Toward a State of Esteem*. Strain and tension were now rife within the group. Seven members dissented from the report, citing "deep irresolvable philosophical and theological differences" and a failure to "recognize the eternal God as the origin of all human worth." One member, yoga teacher David Shannahoff-Khalsa, attacked the group for promoting self-esteem as a "universal remedy for a society that has become increasingly complex and dysfunctional." The report's reliance on "anecdotal wisdom" rather than firm scientific evidence was a more fundamental criticism.[21]

Nevertheless, from the report's opening pages, Vasconcellos portrayed the Task Force's work as having already conferred huge social benefits. In its preface, Andrew Mecca described the project as a "bipartisan pioneering effort to reframe social problem solving" for the "greater wellbeing of the individual and society." He maintained that "self-esteem may well be the unifying concept to reframe American problem-solving" through empowerment and nurturance. Responding directly to his critics, Mecca insisted:

> Self-esteem is not some new age, feel good approach. Rather, it is part and parcel of personal and social responsibility. Governments and experts cannot fix these problems for us. It is only when each of us recognize our individual personal and social responsibility to be part of the solution that we also realize higher self-esteem.[22]

Perhaps the most radical part of the report came from Vasconcellos, who claimed:

> As we approach the twenty-first century, we human beings now, for the first time ever, have it within our power to truly improve our human condition. We can proceed to develop a *social vaccine*. We can outgrow our past failures; our lives of crime and violence, alcohol and drug abuse, premature pregnancy, child abuse, chronic dependence on welfare and educational failure.[23]

The most startling part of the report was probably Vasconcellos's announcement that the Task Force's work rivaled in significance the founding of America and the 1969 moon landings. Vasconcellos insisted that, "in the 1990s, we humans have the opportunity to enter our own inner space. We can unlock the secrets of healthy human development ... with this report, we point the way toward a successful effort to truly improve our human condition." He compared his own vision and "dreams" to those of America's founders in 1776 and John F. Kennedy as he planned to place a man on the moon. Presumably, invoking Kennedy's optimistic dreams of landing a man on the moon was intended to assuage critics who considered Vasconcellos's visions equally unlikely. "Stretch your imagination into the future," he implored, "beyond presently accepted limits."[24]

Vasconcellos's audacious social vision suggests that he himself was far from lacking in self-esteem. His final call to arms in the report read:

> The success of our endeavor now depends on each of us. Will we recognize the centrality of self-esteem in our lives? Will we commit ourselves to developing our own self-esteem and modelling it in our relationships with every other person in our lives? It is a cause well worth our personal commitment and participation. As a beginning, carefully read and digest this report. Incorporate its recommendations in your daily life and all of your relationships. Together, we can truly grow more self-esteem and responsibility. Together, we can truly make a difference. Together, we can make history better. Let us begin right now!

The remainder of the report outlined the Task Force's practical plans. Like much mainstream self-help advice, it depicted families as the "incubator" of self-esteem and childhood as a pivotal time for personality development. Healthy growth depended on parents also possessing self-esteem. Schools and workplaces needed reorganizing around self-esteem principles. In turn, society would be shielded from teenage pregnancy, child abuse, alcohol and drug addiction, violence, and crime. The Task Force had an additional list of social problems that it hoped to tackle in the future: aging, AIDS, homelessness, disability, and racism. The latter was particularly notable for its omission in the first stage, and was suggestive of Vasconcellos's inability to fully acknowledge the diversity of Californian society. Vasconcellos held a utopian vision of a reformed American society entirely transformed simply by everyone feeling better about themselves. By the twenty-first century, perhaps self-esteem would permeate all areas of life, with governments actively empowering citizens to become self-realizing.[25]

A critical problem then arose. Despite their hard work, the university researchers had been unable to decisively identify any causal relationships between low self-esteem and the numerous social problems for which it was blamed. Not even one. Quite apologetically, the report announced: "Just because the causality chain is incomplete with regard to self-esteem, it does not mean that it is implausible." Despite lack of proof, the report proceeded to make sweeping recommendations including developing state-wide media campaigns to raise parental awareness of self-esteem's importance, reshaping school curricula, designing self-esteem initiatives for drug and alcohol addicts, establishing a juvenile justice system based around ideals of personal responsibility, improving the self-esteem of welfare recipients, changing personnel policies in workplaces, and so on.[26] Vasconcellos announced the report's findings some time before the publication of *The Social Importance of Self-Esteem*, a collection of scholarly research essays intended to collate the researchers' findings. Perhaps he hoped that the public was unlikely to read a scholarly essay collection, and would overlook its absence of compelling evidence about self-esteem's apparent societal importance. If so, he was very wrong. International interest was high. Critics had a field day.[27]

From its introductory pages, *The Social Importance of Self-Esteem* admitted that self-esteem was elusive, difficult to measure, and impossible to correlate with specific social problems. Despite these significant methodological obstacles, the editors — Mecca, Smelser, and Vasconcellos — nonetheless highlighted self-esteem's pivotal importance to understanding personal behavior. Neil J. Smesler wrote: "The guidelines for intervention and policies aimed at rehabilitation will be as murky as the state of our knowledge of the phenomena concerned," adding that "wisdom appears to suggest caution, then, in the area of policy."[28] A chapter on child abuse and self-esteem woefully concluded: "We find that we can draw only one firm conclusion: that considerably more knowledge is needed to guide development of social policy and programs to deal with problems of self-esteem." A further chapter lamented: "It will be very difficult indeed to identify a causal link between self-esteem and teenage pregnancy." Linking drug and alcohol misuse to low self-esteem proved equally challenging. "There is a paucity of good research, especially studies that could link the abuse of alcohol and drugs with self-esteem," explained one contributor, adding: "What evidence there is remains inconsistent."[29]

Shannahof-Khalsa later claimed that Vasconcellos deliberately tried to brush these findings under the carpet by delaying the publication of *The Social Importance of Self-Esteem*. He stated: "It was hard to believe that Vasco's task force had been so rash as simply to invent the quote, the one that stated the findings were 'positive and compelling'." Shannahof-Khalsa also claimed that Vasconcellos had threatened to wield his political power to slash the University of California's budget if the research was not undertaken.[30]

Critical perspectives

Despite its fundamental problems, more than 60,000 copies of Vasconcellos's report were distributed in the early 1990s. Positive thinking accelerated its spread across American school curricula. In 1990, the Governor of Maryland established a panel to tackle drug abuse, teenage pregnancy, school failure, and other social problems, based on ideas about making young people feel better about themselves.[31] Nearly all of California's fifty-eight counties established self-esteem forces. Governor Pete Wilson publicly endorsed the movement and declared February 1991 to be "Self-Esteem Month." By 1994, thirty states across America had enacted 170 statutes that sought to promote, protect, or enhance self-esteem.[32] Eventually, the National Association for Self-Esteem replaced California's initial Task Force.

Nonetheless, some teachers remained skeptical. For them, handing out constant praise, especially to grossly underperforming pupils, seemed meaningless. In 1990, Richard R. Doremus, Superintendent of Schools, complained to the *New York Times* that many people were now "under the delusion that all one needs to do to improve self-esteem is to plug in a canned program for one hour a week and thereby reduce dropout rates, reduce drugs and alcohol abuse among school children, help children from broken homes, and address the AIDS crisis. No evidence is provided to support these extravagant claims." Doremus saw some value in short self-esteem courses, but drew the line at basing school philosophies entirely around the vaguely defined concept. A letter was printed in the same issue of the newspaper describing school-based self-esteem initiatives as a "band aid approach to children's wellbeing," a further critique of self-esteem theory's inability to reach the heart of complex social problems.[33]

In 1993, professor of education Lillian G. Katz argued, again in the *New York Times*, that "to pretend that all children are created equal in academic, artistic or athletic ability is to undermine them later in life. Instead, learning to deal with setbacks and maintaining the persistence and optimism necessary for a childhood's long and gradual road to mastery are the foundations of lasting self-esteem."[34] Five years later, Richard F. Elmore, a professor at Harvard, added more forcefully that "for most teachers, self-esteem is a theory they invent to cover the fact that they have low expectations for kids."[35] Even in a relatively supportive 1992 article, *Newsweek* reported that self-esteem "is a matter less of scientific pedagogy than of faith, faith that positive thoughts can make manifest the inherent goodness in anyone, even ten-year-old boys."[36]

Religious critics joined the debates. They were especially concerned with the replacement of God as a central social authority in the individualistic self-help techniques being promoted in schools. It was an innocent-looking blue dragon (Pumsy) and a hand puppet (Duso the Dolphin) that truly raised their ire. Pumsy in Pursuit of Excellence was first introduced in the 1980s, and selected by the Task Force as a model educational program. Opponents angrily claimed that children were being exposed to Eastern religion, the occult, New Age spiritualism, and hypnotic relaxation techniques.

In 1993, Alabama psychiatrist George Twente contended that "putting a child in an altered state and suggesting reliance on an inner authority undermines the child's respect for external authorities like parents and teachers."[37] After one school banned the use of Pumsy in response to a parental complaint, Twente also commented that the puppet was hypnotic and might endanger children. "You're almost convincing the child he has magical powers," insisted Twente, who equated Pumsy with the New Age movement.[38] In July 1992, eight parents at Cedar Wood elementary schools filed a lawsuit asking the district to ban the use of Pumsy in all schools due to its perceived association with psychotherapy, hypnosis, meditation, guided imagery, and use of repetitive self-affirming statements. Once again, Twente publicly criticized the material for encouraging children to make decisions on their own instead of relying on adult guidance.[39]

These claims highlight the problematic relationship between Christianity and the psychological sciences. For conservative Christians, the basic premise of self-esteem presented problems. At its essence, the self-esteem movement advocated worshipping oneself, not God. For that reason, it could be interpreted as more akin to paganism than to Christianity. For many Christian denominations, self-denial was more important than self-worth. Moreover, many Christians viewed psychology as a belief system, not an empirical science. The idea of valuing the opinions of "significant others" was fundamentally pivotal to self-esteem theory, but, for Christians, God was the only "significant other" whose opinions mattered.[40]

In 1993, Jill Anderson, who developed the Pumsy program, begrudgingly rewrote her curriculum to quieten conservative Christians, whom she considered ridiculous. In an interview, she stated: 'They complained that we were teaching children to be schizophrenic. So in the new edition we simply added the explicit clarification: Of course, Pumsy doesn't have three minds." She added: "I have found them by and large to be totally unfounded and reactionary. But if you can reduce conflict in a community by making a few simple changes, why wouldn't you do that?" In a live national television broadcast, author Eric Buehrer criticized Pumsy for teaching children to think for themselves. Anderson retorted "I'm very proud of that."[41] This debate fed into a more general critique of the self-esteem movement: its individualistic dependence upon the self as the basis of social and moral authority. Religious conservatives were threatened by activities that resembled nonorthodox, alternatives systems of faith, and recognized that self-esteem encouraged retreat into the self. Self-esteem practices were secular in nature, and privileged independence over group cohesion.

Phyllis Schlafly, President of the conservative Eagle Forum, was a particularly vocal critic. In 1990, she insisted in a *New York Times* interview: "Instead of teaching them how to read, they're teaching kids to feel good about being illiterate." At the time, the Eagle Forum was monitoring (what its members considered to be) the incursion into schools of a vague psychological concept. It offered support to parents opposed to self-esteem programs and plans for how to confront teachers and school boards. Of course, the great irony here was that self-esteem-based education in fact tacitly supported many of the core values of conservative, white America with which Schlafly herself identified, most notably the traditional, self-contained, nuclear family. Nonetheless, Schlafly emphasized that families were the most appropriate place to inculcate self-esteem. School-based programs offended "traditional" familial values and violated parents' rights to freedom of religion.

Once again, the emphasis on non-Christian psychological techniques proved threatening. Schlafly was notably contemptuous of a self-esteem curriculum recently withdrawn from a Nebraskan school after parents had protested against its use of guided imagery, relaxation, and affirmation, viewing these as indoctrination into New Age and Hindu religions.[42] These critics missed the overarching point that self-esteem school provision, venerating God, or retreating into the nuclear family were all unlikely to do much to resolve society's overwhelming problems.

Meanwhile, in universities and psychological research laboratories, academic researchers continued in their abject failure to gather compelling evidence about low self-esteem's apparent connections with societal ills. As the 1990s progressed, a growing number of skeptics warned that school-based self-esteem initiatives failed to recognize the complexities of young peoples' lives. As one 1997 research article concluded: "Instead of simplistic and naive projects focusing on self-esteem, future research with young people needs to be more sympathetic to the relationships between behavior, context, and meaning, and also needs to involve a recognition of the importance to young people of the varied social environments within which they find themselves."[43]

By the decade's end, some academic researchers openly admitted that, despite more than a century of research, fundamental questions about self-esteem remained unanswered: Why is self-esteem important? Do humans really need self-esteem? Does enhancing self-esteem really reduce social problems, as the self-esteem movement claims?[44] If psychologists failed to gather convincing evidence, then perhaps it was the case that self-esteem was simply a "cultural belief," as argued at the time by education professor Joseph Kahne, seized upon by politicians and educationalists to suit their own needs and ends.[45]

Criminals, gangstas, and classroom avengers

In the 1990s, it became apparent that not all criminals lacked self-esteem. Indeed, supremely confident icons – e.g. Dr. Dre, Snoop Doggy Dogg – seemed more interested in conveying details of their gangsta lifestyle and its violence. This contrasted sharply with ideas that people turned to crime and vice because of their low self-esteem. On the contrary, it seemed possible to both have high self-esteem and enjoyably pursue a criminal lifestyle. This was particularly visible in hip-hop culture, which, by the 1990s, had largely abandoned social activism to instead explore, according to its critics, the perverse pleasures of women, drugs, and shooting one another. At the time, a harmful drug culture was infiltrating many communities, particularly poorer African American ones. Imprisonment had replaced slavery, lynching, and more overt displays of racism as the key means of segregating blacks and whites, so some claimed. Such communities presumably found little of relevance in therapeutic strategies underpinned by white, conservative values.[46]

In the late 1990s, some well-publicized academic articles had challenged Vasconcellos's basic presumptions. Roy F. Baumeister and Laura Smart published a scientific article about "the dark side of high self-esteem."[47] In 1998, Brad J. Bushman and Baumeister refuted ideas that low self-esteem led to crime and violence. Despite decades of research, conclusive evidence on this had still not emerged. Bushman and Baumeister warned that many aggressive people actually held themselves in very high regard, their violence and aggression being spurred on by a deeply ingrained narcissism.[48] Discussing this research, *Newsweek* argued that the agenda of improving everyone's self-esteem was not simply a silly project: it was potentially dangerous. Too much self-esteem encouraged violence, hostility, and aggression. Worryingly, American schools had been boosting everyone's self-esteem since the 1970s.

Bushman and Baumeister's article garnered considerable attention since it followed in the wake of a series of American school shootings. One of these, at Springfield, Oregon, left two students dead and twenty-five wounded. *Newsweek* focused on Luke Woodham who, in 1997 in Pearl, Mississippi, had killed two students and injured seven others after murdering his own mother. Woodman perfectly fitted Bushman and Baumeister's profile of youngsters who possessed far too much self-esteem. In a court-ordered evaluation, three psychologists confirmed that Woodham's narcissistic traits had been brought on by an unstable, unjustified level of self-esteem. He was supersensitive to criticism and slights due to the fragile psychological foundations of his self-esteem. Here was a profoundly narcissistic person who, in responses to criticism, turned astonishingly aggressive. Psychiatrist James Gilligan criticized schools and parents for building up the wrong sort of self-esteem. He said: "There are well-meaning parents who have seen self-esteem as 'every little thing your kid does, praise them to the sky.' But if teaching self-esteem is done wrong, you can raise a generation of kids who cannot tolerate frustration."[49]

As school violence continued to afflict America, discussion ensued about what might be causing this wave of youthful violence. Some international critics blamed America's education focus on self-esteem for having produced morbidly narcissistic tendencies and overblown egos. In 1999, English author Bernard Cornwell wrote an article entitled "In America, They Even Groom their Kids to Kill." This followed the Columbine High School massacre on April 20, which saw ten students and a teacher murdered followed by the suicides of perpetrators Eric Harris and Dylan Klebold. A further twenty-one people were injured and the pair had exchanged gunfire with the police. Causing even further alarm, the attack also involved several homemade bombs, the largest of which had been placed in the cafeteria and, had it detonated, would undoubtedly have killed all students eating in there. Other car bombs were placed in the parking lot and on the road leading to the school, targeted at first responders. At the time, this was the deadliest shooting at an American school and inspired various copycat incidences. The motives still remain unclear.

Seeking explanation, Cornwell observed that, "while American students rank near the bottom in international attainment tests, their self-esteem is the highest in the world. They're dumb but they believe they're wonderful. This belief comes from the American culture which continually claims that children are special." He continued by stating:

> Children are encouraged to believe that all of society, all of politics, even the bombers over Kosovo, exist just for them … this encouragement of narcissism began, like most insanities, in California, where a director of education

promulgated the idea that children were like sunflowers. If you left them alone, they would grow tall and strong towards the sun; if you tied them down they would be warped and stunted.

Cornwell concluded that it was unsurprising that "adolescent males excluded from the Baywatch world of beautiful jailbait sometimes run amok. Guns are readily available and the discipline with which society used to protect itself against the stupid brutalities of young males has been stripped away in the name of self-esteem."[50]

In subsequent decades, speculation emerged that most "classroom avengers" in fact suffered low self-esteem, and that experiences of being bullied might manifest itself in violent revenge.[51] Nonetheless, Bushman persisted with his claims that mass school shootings were complex and that perpetrators usually had large egos and narcissistic tendencies. Most sought fame.[52] Inculcating self-esteem in schools, some claimed, had created a generation of overly confident criminals whose ideas had infiltrated popular culture (most notably in violent hip-hop lyrics) and became evident in a wave of school shootings.

Self-esteem's ascendancy was on the wane. Its popularity had peaked in the strident efforts to cure California's social ills in the late 1980s, but this fragile optimism regarding the potential of self-help to tackle so many societal problems had ended up exposing the many limitations inherent in using introspective techniques for grandiose purposes. Combined with the additional scrutiny being awarded to feminist discussion of self-esteem, it became apparent to many that the massive faith being placed in self-esteem building was unwarranted. Self-esteem remained notoriously difficult to define and measure, and its precise relation to the social problems that its advocates set out to resolve proved notoriously difficult, maybe even impossible, to prove. In the decades that followed, self-esteem's critics mounted. But perhaps we have much to learn from the mistakes made in the late 1980s. Maybe if individuals such as Vasconcellos had paid more attention to the specific emotional needs of Black, gay, and other marginalized communities, a more inclusive program might have been developed. And if more thought had been given to the potential political overtones rife in self-help strategies, then maybe some of these could have been countered or attended to.

NINE

The Twenty-first Century

It should now seem clear that our feelings of self-esteem, whether high or low, aren't simply some natural, integral part of our psychological makeup that has changed little over time. Instead, self-esteem is a feeling, a psychological state, that is historically and socioculturally contingent. My focus on postwar America has illuminated a remarkably important context in which self-esteem, as an idea, flourished largely due to that country's individualistic ethos bolstered by the increasingly pervasive values of capitalism, conservatism, and neoliberalism. I've followed trends in sociology, philosophy, and histories of emotions and medicine which refuse to believe that a stable, authentic, "true" self exists. Instead, the "self" is a product of discourse and representational symbols unique to specific historical and sociocultural contexts.[1]

Much of this book has focused on the growing influence of a particular interpretation of self-esteem that became dominant in self-help literature, child development literature aimed at parents and teachers, and school-based educational strategies, all of which exhibited a fascination, even an obsession, with improving self-esteem levels. Modern ideas of self-esteem emerged around the mid-twentieth century as a more positive departure from Freudian narcissism and Adlerian inferiority complexes. During the counterculture and beyond, humanistic psychologists saw self-esteem as pivotal to human growth and self-realization. Unlike narcissism, this type of self-esteem was not to be restrained or treated with caution. Self-esteem levels were seemingly boundless and largely positive. Narcissistic problems tended to arise only if positive self-feelings were not authentic or truly ingrained, if they were a façade masking negative emotions.

Self-esteem, like self-help in general, was fundamentally political. As the postwar period progressed, self-esteem practices looked suspiciously like a tool of rightwing conservatism and, from around the 1970s, neoliberalism (although we should note, with some irony, that many rightwing and religious critics perceived psychological self-help with distrust). Self-esteem literature written for parents explored and upheld the primacy of the (nuclear, middle-class) family as the locus of morals and appropriate socialization. Literature aimed at teachers encouraged educational approaches that transferred blame for the problems of social modernity onto individuals rather than adverse socioemotional environments. Teenagers were instructed to work on their self-esteem to help conform, not rebel against, postwar American democratic principles. In short, it shared much in common with currently popular emotions-based strategies such as wellness and mindfulness, which have been similarly criticized for having little to offer those seeking social change and the redress of injustice.

Nonetheless, I have self-consciously avoided depicting self-esteem entirely as a malign, monolithic, insidious concept that has duped us all into accepting the status quo and dissuaded us from collective action against the structural forces that take their toll on our mental wellbeing (although I do consider such perspectives important). Instead, the self-esteem concept is nuanced and multidimensional. It can appeal to various types of politics, even seemingly incompatible ones. Self-esteem discourses emerged from "below," as well as "above." Notably, in the political activism of African American, gay and lesbian, and feminist activists, self-esteem was weaponized. It provided a means of expressing and communicating specific emotional experiences felt across marginalized communities. It offered a means of expression for individuals who failed to identify with, or see much personal relevance in, mainstream self-help literature. The latter variety of self-help literature largely ignored activist approaches, presenting instead problematically generalized styles of self-help advice that perpetuated the idealization of white, straight, middle-class American values, but activist groups remained aware of all of this and produced their own, competing varieties of self-help, which coexisted alongside the dominant self-help narrative.

In developing a fuller understanding of self-esteem's complex history, I hope to promote deeper reflection on the nature of therapeutic culture, its positives, negatives, and ambiguities. In the twenty-first century, self-esteem might feel a little eclipsed by wellness and mindfulness, but it still remains an important emotional pursuit. The amount of popular and scholarly material still being published on the subject is staggering. One author, Franklin Holloway, opened his book with the statement: "Self-esteem construct has spawned a research literature of such magnitude and richness that it is impossible to summarize."[2] Despite this, as the first quarter of this century draws to a close, the self-esteem concept remains theoretically vague, flaky, and uncertain. Major gaps in knowledge endure. Researchers approach self-esteem in contrasting ways, and using different terminologies. Basic concepts such as self-worth are used interchangeably with self-esteem. Fundamental questions still remain unanswered. For instance, how can we separate negative people's attitudes toward themselves from their general negativity? What comes first: low self-esteem or underperformance?[3]

Chapter 1 explored the rising popularity of the self-esteem concept in professional and public circles early in the twentieth century, culminating in the late 1960s. Given that we've now been urged to work on our self-esteem for a century (at least), one might presume that we would have learnt to feel happier about ourselves by now. This isn't the case. Without denying that self-help approaches might well have helped some people feel better on an individual level, a discernible trend toward general happiness appears not to have occurred. Perhaps this is because the harmful socioemotional environments that impede our happiness remain in place. Maybe self-help generally doesn't work. While debates about American narcissism rage on, self-esteem literature differs in the sense that it actively discourages everyone from being ordinary.

Psychotherapist Polly Young-Eisendrath describes the "self-esteem trap," the processes by which encouraging everyone to be special, a winner, only promotes dissatisfaction with lives that are generally fine. A certain amount of self-esteem is desirable and healthy, but does everyone really need to feel special and extraordinary? In such ways, self-focus harmfully produces shame, anxiety, pressure. Inevitable failures lower our self-esteem further still.[4] Nonetheless, self-help literature, with Vex King's 2018 *Good Vibes, Good Life* providing just one of many examples, continue to criticize readers for being "comfortable with mediocrity" and avoiding "a greater life, one that's beyond what most consider the norm."[5] These are the inherent contradictions in the all-important pursuit of happiness.

Some consider ongoing issues in self-esteem management to be related to a lack of thorough knowledge among psychologists and counselors, who think they know what self-esteem is, but in reality haven't thought closely enough about it. Counseling professor Mary H. Guindon complains: "So many practitioners are not particularly knowledgeable about self-esteem, although they tell me that low self-esteem plagues their clients." She argues that therapeutic applications of self-esteem theory need to be underpinned with a fuller theoretical understanding of the subject.[6] But there are bigger forces at play. I agree with scholars such as sociologist Eva Illouz, who claim that Western modernity has brought with it a vast amount of destruction, and that its key values, including individualism, remain in place with no better alternative in sight.[7] However, in addition to acknowledging the emotional impacts of large forces such as capitalism, self-esteem literature could learn from, and benefit from including, more specific contexts of self-esteem experience. Much self-help literature still perpetuates a blindness toward those outside the "dominant culture." For such reasons, much of this book has criticized self-esteem culture for ignoring the specific needs of marginalized communities. Nonetheless, Chapters 2 and 3 explored how Black and LGBT activists nonetheless developed their own self-help initiatives, seeing little relevance in mainstream self-help literature that essentially informed them to live up to American ideals of being white, straight, and middle class.

Despite decades of African American activism and persuasive work by (usually Black) social psychologists on the emotional fallout of racism and discrimination, most self-help literature still refrains from engaging fully with these problems. What literature exists sometimes takes a political stance, pointing to Black people's long American history of slavery and discrimination.[8] Others focus more on immediate day-to-day problems. Researchers remain as divided as ever on whether Black students tend to underperform at school due to their pathologically low self-esteem or because they still can't fully identify with an educational system believed not to be fully invested in their scholarly success. Debates continue about whether Black self-esteem is a "myth" or whether discriminated-against communities enjoy surprisingly high levels of psychological resilience. Recent anthologies of Black writing, such as Tarana Burke and Brené Brown's *You Are the Best Thing* (2021), still foreground lingering shame and emotional vulnerability.[9]

This may not simply be a white versus black issue. Undoubtedly, in the 1960s, many Black people felt invigorated by the emphasis on improving emotional wellbeing. African American minister, activist, and author Michael Eric Dyson fondly remembered the 1960s call for Black pride for its positive psychological effects. He comments: "The stakes of self-regard are greatly increased when a healthy pride has been systematically denied an entire group of people, because of race, gender, sexual identity, and the like."[10] However, bell hooks accused Black leaders, since the 1960s, of losing interest in the need for Black psychological development once some civil rights and economic gains had been made. Writing from a feminist perspective, hooks attributed this partly to the lack of female leadership, which might have created space for holistic self-development. hooks believed that the minds and imaginations of Blacks were left colonized, as the 1960s activists never completed their psychological projects.[11]

Other Black authors recognize the more general forces of capitalism that affect us all, but point to the unique emotional experiences of their communities. For instance, in 2016, Eddie S. Glaude Jr. insightfully outlined the ludicrousness of presenting "Black misery" as being caused by individual behavior and bad choices, essentially as a private affair. In his sardonic words, "somehow people believe, and they have done so for much of our history, that Black social misery is the result of hundreds of thousands of unrelated bad individual decisions by Black people all across this country."[12] Glaude Jr. recognizes the problem as running far deeper. Meanwhile, Keeanga-Yamahtta Taylor has recently argued that blaming Black people for inequality and irresponsible behavior is a process not always about race, or even white supremacy, but about making sense of, and rationalizing, poverty in ways that absolves the state of any culpability.[13] Black conservatives go so far as to argue that white racism initially created feelings of inadequacy in the past, but Black people had since perpetuated those feelings. According to this line of argument, Black and white people alike suffer from low self-esteem as part of the twentieth- (and twenty-first-) century human condition, but Black people have nurtured a sense of victimhood around this. Here we see, ironically, Black authors encouraging Black readers to forget about their specific problems and barriers, about the social injustices facing them, about racism.[14]

If self-help were to shed its color-blindness, perhaps more culturally specific programs could be created that truly address all Americans. Perhaps, too, self-help literature could integrate the collective, rather than individualistic, perspectives that have proven so central to Black activism on emotional and mental health issues. We need also to take intersectionality into account, and ensure that adequate attention is paid to the emotional wellbeing of African American girls, and also Black lesbian and gay people, who simultaneously endure multiple, overlapping forms of discrimination and social disadvantage.[15] As things stand, discussion of African American self-esteem sometimes appears briefly in self-help literature, usually feeling more like a token gesture. Black babies cynically appear on the front covers of self-esteem literature aimed at parents, but rarely feature much on the inner pages.

Emotional issues relating to sexuality are in fact usually discussed in self-esteem books aimed at teenagers, but adults still rely, for the most part, on their own specialist genres of self-help, building on the historical developments outlined in Chapter 3. The newest additions aim to help transgender communities with their emotions. As with much activist self-help literature in the past, many of these are marketed in terms of "survival" in a hostile world.[16] LGBT psychologists still argue that theory, research, and clinical practice still fails to address intersectional realities. More could still be done to address the salience of race, ethnicity, gender, gender expression, disability, and social class in LGBT psychology. As with Black and feminist research, it seems that much of this has been driven by LGBT psychologists themselves, rather than having been fully integrated into psychology.[17] Shame still figures prominently in some LGBTQ+ literature, articulating a certain sanguineness about gay pride. Walt Odets writes that the mere assertion of gay pride does not undo shame, it masks it.[18] Therapist Alan Downs writes that "as the eye cannot see itself, we [gay men] cannot see or feel this embedded shame. But make no mistake, the shame is there and it is very real."[19]

Having established the multiple political uses surrounding self-esteem management, this book has traced how the self-esteem concept simultaneously became entwined with neoliberal ideologies that sought to sustain, and promote adjustment to, Western democratic values, while also exhibiting a blindness to matters such as race and sexuality. It has achieved this largely by choosing to ignore the various consciousness-raising activities that have been so prominent in counter-culture, Black, and LGBTQ+ activism. Through various kinds of self-help literature, parents, teachers, youngsters, and adults were exposed to images of the idealized, healthy American, brimming with self-esteem and happiness. Typically, this individual resembled a white, straight, middle-class ideal, individuals fully invested in Western democracy with its values of individualism and consumerism.

The middle-class ideal of offering infants an emotionally expressive world based around family values discussed in Chapter 4 persists in the twenty-first century. Websites such as Psychology Today and Ask Dr. Sears still provide much the same advice as in earlier decades. Smile at your baby often. Talk to her, sing to her, read to her. Praise your baby as she burps. Apologize if being too harsh. All of this will help raise a cooperative, happy child.[20] Parents themselves are still urged to learn self-respect and gain confidence to maintain familial harmony. School performance, work achievements, marital success, happiness: all of this might one day be in jeopardy should you fail to instill a positive, self-image.[21] This is essentially the message transmitted in such sources, underpinned by the morally persuasive message: Who wouldn't want a happy child?

Parent-blaming (usually directed toward the mother) continues to feature strongly in self-esteem guidance aimed at parents that still transmits messages about motherhood and domesticity that overlook life's diversities. One example is pediatric physical therapist Tara Losquadro Liddle's *Why Motor Skills Matter* (2004). Liddle lambasts parents who do not seek professional help about their child's development, the result being, as she puts it, "well-meaning but misguided parenting."[22] In such statements, one finds echoes of the longstanding tradition of parent-blaming, combined with an emphasis on time-consuming, emotionally intensive motherhood that only families with time on their hands could fully engage with. Mothers are encouraged to think about themselves as anxious, insecure, or confused individuals who need to become confident mothers. The continued cutting back of welfare, social security, and public infrastructures in the twenty-first century is currently intensifying this privatization of childraising and the responsibilities placed upon mothers. Meanwhile, middle-class parents, particularly mothers, remain positioned as role models of quality parenting for everyone else to emulate.[23]

A similar situation arises in self-help literature aimed at parents of teenagers, which typically warns of a dramatic self-esteem decline around age 12, which worsens through to adulthood. Moira McCarthy depicts the teenage daughter, hindered by low self-esteem, as hysterical and angry due to the psychological insecurities bubbling under the surface. Only by developing emotional maturity, including acquiring self-esteem, can she become a calm individual. Ironically, such perspectives perpetuate misogynistic caricatures about female emotional life.[24] Others list depression, teen pregnancy, drug use, and crime as the social problems that result from teenage girls' flailing levels of self-esteem. The solution? Tell your daughter she's pretty. Comment positively on her figure. Encourage her to join the Girl Scouts.[25] At worst, some self-help literature warns readers that they might have been emotionally abused, neglected, or smothered by their parents while growing up without even realizing it, the consequences of which include self-destructiveness, poor self-image, depression, and suicidal thoughts.[26]

As Chapters 5 and 6 outlined, self-esteem made one of its strongest impressions in the postwar American school system. Since then, educational psychologists have borne the brunt of self-esteem's critics. Well into the 2010s, many educators uphold high self-esteem, among both teachers and pupils, as crucially important to a rounded education.[27] Not everyone is convinced. Even defenders of self-esteem-based education concede that some of their fundamental presumptions remain unsupported by decisive evidence.[28] However, the basic idea that educational environments that support emotional learning are useful remain in place. EQ is as important as IQ.[29] New approaches have emerged such as social-emotional learning (SEL), which, as the title suggests, emphasizes emotions. In 2019, a report emerged entitled *From a Nation at Risk to a Nation at Hope*, which likewise promoted developing social, emotional, *and* academic skills in schools to help produce "caring and competent" adults. The study discussed students of color, though only briefly. No serious discussion took place of their emotional makeup and its impacts upon educational achievement.[30] Now, as in the recent past, serious exploration of the additional challenges faced by Black youth in securing self-esteem remains largely confined to specialist research on racial socialization or Black literature, rather than being fully embedded into educational approaches.[31]

Women and girls provide a notable exception to the rule of ignoring the specific needs of certain social groups in self-esteem literature, perhaps largely because of the attention drawn to female self-esteem by leading feminists from the 1980s and 1990s. Nonetheless, as Chapter 7 recognized, the genre of female-focused self-esteem literature is replete with problems. The Strong Girls program proved popular in California in the 2010s. This resuscitated arguments that girls suffer lower self-esteem than boys, despite excelling academically. Like much feminist literature, this program tackles serious problems such as rape culture, familial abuse, suicide, the gender pay gap, and also problems such as girl-on-girl bullying.[32] All of these are undoubtedly worthy causes, and ones that focus on the needs of a specific group. But once again we enter the vicious circle. How exactly can building self-esteem on a personal level really address broader social inequalities? Are we still unwittingly teaching girls how to adjust to, and cope with, life's barriers instead of tackling them collectively head-on? And is there a risk again of pathologizing teenage girls' lives by presenting them as full of horrors and dangers, often in ways that not all girls identify with?

That said, there does exist a more uplifting genre of self-help literature aimed at women and teenage girls, with fun titles such as *The Confidence Code for Girls: Taking Risks, Messing Up and Becoming Your Amazingly Imperfect, Totally Powerful Self* and *Girls without Limits: Helping Girls Succeed in Relationships, Academics, Careers, and Life*.[33] Without denying that women face serious barriers rooted in patriarchal systems, these books resonate with female self-help readers who want to feel good, sassy, and smart. These texts have considerable appeal, adopting a different approach than feminist-based texts. Teenagers continued to be blamed for more or less everything, but it's equally encouraging that a less scaremongering genre of self-esteem literature is being written for them, often focusing on positive matters such as body image, deciding which social justice issues to care about, feeling good, and getting along with friends, parents, and teachers. Whereas in the 1980s much self-esteem literature was written for parents, instilling fear about what their teenagers might be up to, advice aimed directly at teenagers tends to adopt a more upbeat tone, even when addressing issues of a more serious nature.[34]

Nonetheless, in 2022, Shani Orgad and Rosalind Gill lamented that the "confidence culture" only accelerated after around the early 2010s, and is more often than not targeted directly at women. Usually through online technologies such as apps, women are encouraged to expose their vulnerabilities (and, indeed, to think of themselves as vulnerable). In Orgad and Gill's words, vulnerability "has become almost mandatory and authorizes the individualistic psychologized confidence imperative." They take issue not so much with the idea of feeling confident and full of self-esteem, but with the cultural prominence of such ideas and the ways in which feminism itself still ironically props up the "confidence culture."[35]

Problematically too, self-esteem-enhancing therapies often remain upheld as a panacea for all manner of social problems, while also being blamed for producing self-obsessed, narcissistic individuals, a development that perhaps reached its peak in the inauguration of the Task Force to Promote Self-Esteem and Personal and Social Responsibility in late-1980s California, discussed in Chapter 8. People aren't confident and happy. John Vasconcellos's dreams never came true. Some self-esteem books adopt a fun, upbeat approach. We now have titles such as *The Complete Idiot's Guide to Enhancing Self-Esteem* and *Self-Esteem for Dummies*.[36] But the very worst examples are potentially damaging. An often-reprinted example is Gael Lindenfeld's *Self-Esteem: Simple Steps to Build Your Confidence*, which has sold more than 1 million copies worldwide. Lindenfeld claims that when she started writing the book in 1994, the public were largely unaware of the idea of self-esteem. (The chronology outlined here would suggest otherwise.) In the third edition, Lindenfeld details how working on her self-esteem helped her get over the death of a daughter.[37] Perhaps so. But I would worry about whether such claims discourage people suffering severe grief from seeking more appropriate professional help beyond being encouraged to feel better about themselves. (I should note that some of these books actually do realize the limitations of self-help and promote professional help for problems of a more serious nature, although not all.)[38]

Embrace your greatness! So writes Judith Belmont in her 2019 book. Sure, there's nothing wrong with wanting to feel better about oneself. But problems arise when Belmont, on the very first page, begins to describe low self-esteem as the "number one culprit underlying almost all common life problems." Low self-esteem breeds anxiety, depression, and stress, she writes, while also causing relationship problems, education and work difficulties, social disconnectedness, and loneliness. Maybe it does, but such literature needs to acknowledge that broader social forces are at work that harm our emotional wellbeing, and which require much more than individual action to be addressed. As if all of that wasn't enough, Belmont warns that low self-esteem lowers life expectancy. Those who refuse to heed her advice can therefore expect an earlier death than hoped for.[39]

It seems that self-help literature still lives in the hangover from the postwar period when it was common to publish morally persuasive and fearmongering claims. Twenty-first-century self-esteem literature differs, with its inclusion of cognitive behavioral therapy and mindfulness techniques.[40] These Eastern-inspired approaches would have been actively condemned by the likes of Phyllis Schlafly in the 1980s. Nonetheless, they still carry broadly similar messages about personal responsibility, individual action, the family, and so on. Regrettably, Belmont also touches upon sexual, physical, and emotional abuse before recommending that readers listen to Whitney Houston's 1986 hit "The Greatest Love of All," a song about the importance of self-love.[41]

Alternatives have been found in ideas such as "grit," which focus not on constant praise but on "perseverance and passion for long-term goals." "Grit" involves harder work than simply being praised, but instills a more authentic, deeply rooted mindset. It focuses more on selective praise and on building on small successes in realistic ways.[42] Nonetheless, "grit" winds up being equally moralistic, with a lack of persistence being experienced as personal failure. Researchers question the idea, although many parents seemed to have responded well.[43] Another idea revolves around the "Quiet Ego." Recognizing problems in the late twentieth-century emphasis on giving out trophies to children and having them chant "I'm special," which perpetuates idealized images of the self-confident American personality, this approach focuses on helping children develop genuine, ingrained confidence. The ego's volume is turned down so it listens to others, as well as the self, in ways that are humane and compassionate. In some ways, perhaps we are heading back to Victorian ideals of seeing self-esteem as a feeling that's better contained rather than being an all-consuming aspect of one's personality.[44]

Since the 1990s, "positive psychology" has been promoted by Martin Seligman, a prominent American psychologist, educator, and self-help author. Working with Christopher Peterson, Seligman set out to create a positive counterpart to the *Diagnostic and Statistical Manual of Mental Disorders* (DSM). Instead of focusing on pathological problems, books such as *Character Strengths and Virtues* emphasized what can go right with people's mental health. Seligman downplayed the negatives and stressed the positives. Interestingly, he argued that humanistic psychologists, since around the 1960s, had dangerously emphasized self-esteem while discouraging collective action.[45] Perhaps so. But Daniel Horowitz describes Seligman's "positive psychology" itself as a largely white, middle-upper-class movement mostly lacking explicit reference to issues such as race, class, homosexuality, and sexual ambiguity, and displaying a relative insensitivity to cultural differences. As further evidence, Horowitz comments on the absence of African Americans from positive psychology's leading scholars.[46]

We can't just abandon the idea of self-esteem now we've come this far. Low self-esteem is a significant problem for many people who would benefit from help. This is often overlooked in critiques of capitalist modernity and narcissism, which risks not taking people's negative emotions seriously enough. We could rein in self-esteem's excesses and apply the concept more carefully, shedding any grandiose claims.[47] To do so would require a fuller awareness of how self-esteem is fundamentally political, as this book should hopefully have demonstrated beyond doubt. Yet we would also benefit from paying more attention to the potentially positive aspects of therapeutic culture, its potential to provide a means by which marginalized groups can express and articulate their own emotional experiences, which can differ drastically from those of the dominant culture. And we need to shed our blindness to social injustice issues in self-help literature and listen to the broadest range of voices.

Notes

Self-Esteem: An Introduction

1 Richard Wilkinson and Kate Pickett, *The Inner Level: How More Equal Societies Reduce Stress, Restore Sanity and Improve Everyone's Well-being* (Allen Lane, 2018), 3.
2 E.g. Carol Z. Stearns and Peter Stearns (eds.), *Emotions and Social Change: Towards a New Psychohistory* (Holmes and Meier, 1988).
3 Wilkinson and Pickett, *Inner Level*, 16.
4 James Vernon, *Distant Strangers: How Britain Became Modern* (University of Berkeley Press, 2010).
5 Peter Berger, *The Homeless Mind: Modernization and Consciousness* (Vintage Books, 1974).
6 For sociological perspectives, see Ulrich Beck, *Risk Society: Towards a New Modernity* (SAGE, 1992); Zygmunt Bauman, *The Individualized Society* (Polity, 2001).
7 Fay Bound Alberti, *A Biography of Loneliness: The History of an Emotion* (Oxford University Press, 2021).
8 David Forbes, *Mindfulness and its Discontents: Education, Self and Social Transformation* (Fernwood, 2019), 120; Nikolas Rose and Des Fitzgerald, *The Urban Brain: Mental Health in the Vital City* (Princeton University Press, 2022).
9 Svend Brinkmann, *Diagnostic Cultures: A Cultural Approach to the Pathologization of Modern Life* (Routledge, 2016).
10 Li Zhang, *Anxious China: Inner Revolution and Politics of Psychotherapy* (University of California Press, 2020).
11 Philip Cushman, *Constructing the Self, Constructing America* (Addison-Wesley, 1995), 245–6.
12 David Riesman, with Reuel Denney and Nathan Glazer, *The Lonely Crowd* (Yale University Press, 1950); Norman Vincent Peale, *The Power of Positive Thinking* (Prentice Hall, 1952); Will Storr, *Selfie: How the West became Self-Obsessed* (Picador, 2017), 126.
13 Noted in Rebecca Richardson, *Material Ambitions: Self-Help and Victorian Literature* (Johns Hopkins University Press, 2021), 2.

14. Ellen Herman, *The Romance of American Psychology* (University of California Press, 1995), 270–1.
15. Abraham H. Maslow, *Motivation and Personality* (Harper Row, 1954).
16. Julia A. Boyd, *In the Company of My Sisters: Black Women and Self-Esteem* (Dutton, 1993), 25–6.
17. Jason Schnittker, *Unnerved: Anxiety, Social Change, and the Transformation of Modern Mental Health* (Columbia University Press, 2021), 1.
18. Alain Ehrenberg, *The Weariness of the Self: Diagnosing the History of Depression in the Contemporary Age* (McGill-Queen's University Press, 2015), xxx, 4–5, 8–9.
19. Alberti, *Biography of Loneliness*, 9, 16.
20. Wilkinson and Pickett, *Inner Level*, 19, 4.
21. Examples of critical overviews include Graham Richards, *Putting Psychology in its Place* (Routledge, 1996); Dai Jones and Jonathan Elcock, *History and Theories of Psychology: A Critical Perspective* (Arnold, 2001).
22. Herman, *Romance of American Psychology*, 310; see also James H. Capshew, *Psychologists on the March: Science, Practice and Professional Identity in America, 1929–1969* (Cambridge University Press, 1999).
23. Mathew Thomson, *Psychological Subjects: Identity, Culture and Health in Twentieth-Century Britain* (Oxford University Press, 2006).
24. Nikolas Rose, *Inventing Our Selves: Psychology, Power and Personhood* (Cambridge University Press, 1996).
25. Rose, *Inventing Our Selves*, 22.
26. Michael Eric Dyson, *Pride* (Oxford University Press, 2006), 13.
27. Ronald W. Dworkin, *Artificial Happiness: The Dark Side of the New Happy Class* (Carroll and Graf, 2006), 173, 224.
28. For elaboration, see Charles Taylor, *Sources of the Self: The Making of the Modern Identity* (Harvard University Press, 1992).
29. James L. Nolan Jr., *The Therapeutic State: Justifying Government at Century's End* (New York University Press, 1998), 3.
30. Maureen Stout, *The Feel-Good Curriculum: The Dumbing Down of America's Kids in the Name of Self-Esteem* (Perseus Books, 2000), 15.
31. Paul C. Vitz, *Psychology as Religion: The Cult of Self-Worship*, 2nd ed. (William. B. Eerdmans, 1977), 15–19.
32. Glynn Harrison, *The Big Ego Trip: Finding True Significance in a Culture of Self Esteem* (Inter-Varsity Press, 2013), 18–19.
33. Storr, *Selfie*, 92.
34. L. Edward Wells and Gerald Marwell, *Self-Esteem: Its Conceptualization and Measurement* (SAGE, 1976), 5, 7–8.

35 Morris Rosenberg, *Conceiving the Self* (Basic Books, 1979), 1; Stanley Coopersmith, *The Antecedents of Self-Esteem*, 2nd ed. (Consulting Psychologists Press, 1981), 1, 28.
36 E.g. Fiona King and Adrian Lohan, "Self-Esteem: Defining, Measuring and Promoting an Elusive Concept," *REACH Journal of Special Needs Education in Ireland*, 29 (2016).
37 Peggy J. Miller and Grace E. Cho, *Self-Esteem in Time and Place: How American Families Imagine, Enact and Personalize a Cultural Ideal* (Oxford University Press, 2018), xxii, 82.
38 Frank Furedi, *Therapy Culture: Cultivating Vulnerability in an Uncertain Age* (Routledge, 2004), 4, 159.
39 Allan V. Horwitz and Jerome C. Wakefield, *All We Have to Fear: Psychiatry's Transformation of Natural Anxieties into Mental Disorder* (Oxford University Press, 2012); Allan V. Horwitz, *The Loss of Sadness: How Psychiatry Transformed Normal Sorrow into Depressive Disorder* (Oxford University Press, 2012).
40 Thomas Szasz, *The Myth of Mental Illness* (Hoeber-Harper, 1961).
41 Furedi, *Therapy Culture*, 5–6.
42 Tommy Dickinson, *"Curing Queers": Mental Health and Their Patients, 1935–74* (Manchester University Press, 2016).
43 Kurt Danziger, *Constructing the Subject: Historical Origins of Psychological Research* (Cambridge University Press, 1994).
44 Ian Hacking, "The Looping Effect of Human Kinds," in Dan Sperber, David Premack, and Ann James Premack, *Symposia of the Fyssen Foundation. Causal Cognition: A Multidisciplinary Debate* (Clarendon Press, 1995); *The Social Construction of What?* (Harvard University Press, 1999), 105.
45 Brinkmann, *Diagnostic Cultures*, 37.
46 Rose, *Psychological Complex*; Darrin M. McMahon, *In Pursuit of Happiness: A History from the Greeks to the Present* (Allen Lane, 2006).
47 Storr, *Selfie*, 18–19.
48 Tom Wolfe, "The 'Me' Decade and the Third Great Awakening," *New York* (23 August 1976); Edwin Schur, *The Awareness Trap: Self-Absorption instead of Social Change* (McGraw-Hill, 1976); Christopher Lasch, *Culture of Narcissism: American Life in an Age of Diminishing Expectations* (W.W. Norton, 1979); Robert D. Putnam, *Bowling Alone: The Collapse and Revival of American Community* (Simon and Schuster, 2000); Jean M. Twenge and W. Keith Campbell, *The Narcissism Epidemic: Living in an Age of Entitlement* (Free Press, 2009); Elizabeth Lunbeck, *The Americanization of Narcissism* (Harvard University Press, 2014).
49 Lunbeck, *Americanization of Narcissism*, 109–12.

50 Storr, *Selfie*, 17.
51 Ronald E. Purser, *McMindfulness: How Mindfluness Became the New Capitalist Spirituality* (Watkins Media, 2019), 110.
52 Philip Rieff, *The Triumph of the Therapeutic* (Chatto, 1966).
53 Thomas Szasz, *The Therapeutic State: Psychiatry in the Mirror of Current Events* (Prometheus, 1984).
54 Forbes, *Mindfulness and its Discontents*, 120.
55 Daniel Horowitz, *Happier? The History of a Cultural Movement that Aspired to Transform America* (Oxford University Press, 2017), 94, 224; Furedi, *Therapy Culture*, 93; Rose, *Inventing Our Selves*, 75; Michel Foucault, "Governmentality," in Graham Burchell, Colin Gordon and Peter Miller (eds.), *The Foucault Effect: Studies in Governmentality* (University of Chicago Press, 1991); David Harvey, *A Brief History of Neoliberalism* (Oxford University Press, 2005), 65.
56 Carl Cederström and Andre Spicer, *The Wellness Syndrome* (Polity, 2014); Thomas Joiner, *Mindlessness: The Corruption of Mindfulness in a Culture of Narcissism* (Oxford University Press, 2017); Forbes, *Mindfulness and its Discontents*, 23, 28.
57 Purser, *McMindfulness*, 7–8, 245.
58 Angela Duckworth, *Grit: Why Passion and Resilience are the Secrets to Success* (Vermilion, 2017).
59 Joanna Bourke and Robin May Schott (eds.), *Resilience: Militaries and Militarization* (Palgrave, 2022).
60 Herman, *Romance of American Psychology*, 16; Wilkinson and Pickett, *Inner Level*.
61 Ole Jacob Madsen, *The Therapeutic Turn: How Psychology Altered Western Culture* (Routledge, 2014), 13.
62 Carl Cederström, *The Happiness Fantasy* (Press, 2018), 10–11.
63 Madsen, *Therapeutic Turn*, 13, 20.
64 Sandy McDaniel and Peggy Bielen, *Project Self-Esteem: A Parent Involvement Program for Elementary-Age Children* (B.L. Winch, 1986), vii.
65 Barbara Cruikshank, *The Will to Empower: Democratic Citizens and Other Subjects* (Cornell University Press, 1999), 4.
66 Andrew J. Polsky, *The Rise of the Therapeutic State* (Princeton University Press, 1991), 2–3, 6; Dana L. Cloud, *Control and Consolation in American Culture and Politics: Rhetorics of Therapy* (SAGE, 1998), xv–xiv.
67 Cushman, *Constructing the Self*, 336–7.
68 Katie Wright, *The Rise of the Therapeutic Society: Psychological Knowledge and the Contradictions of Cultural Change* (New Academia Publishing, 2011).

69 Timothy Aubry and Trysh Travis, "Rethinking Therapeutic Culture," in Timothy Aubry and Trysh Travis (eds.), *Rethinking Therapeutic Culture* (University of Chicago Press, 2015), 3.
70 Michael E. Staub, "Radical," in Aubry and Travis, *Rethinking Therapeutic Culture*, 106.
71 bell hooks, *Rock My Soul: Black People and Self-Esteem* (Atria, 2003).
72 Walt Odets, *In the Shadow of the Epidemic: Being HIV-Negative in the Age of AIDS* (Strand, 1995); Walt Odets, *Out of the Shadows: The Psychology of Gay Men's Lives* (Penguin, 2019).
73 Gloria Steinem, *Revolution from Within: A Book of Self-Esteem* (Bloomsbury, 1992).
74 http://www.slate.com/articles/arts/culturebox/2017/04/the_history_of_self_care.html.
75 https://advice.theshineapp.com/articles/how-you-can-honor-the-radical-history-of-self-care.
76 https://www.teenvogue.com/story/the-radical-history-of-self-care.
77 https://www.bbc.co.uk/programmes/articles/5GwXsvJp6q8PM2RLL2Dgc00/the-radical-history-of-self-care.
78 Ibram X. Kendi, *How to Be an Anti-Racist* (Bodley Head, 2019), Chs. 14, 15.
79 Rafia Zakaria, *Against White Feminism* (Hamish Hamilton, 2021).
80 E.g. Julia A. Boyd, *In the Company of My Sisters: Black Women and Self-Esteem* (Penguin, 1997).
81 Julian B. Carter, *The Heart of Whiteness: Normal Sexuality and Race in America, 1880–1940* (Duke University Press, 2007).

Chapter 1: Happiness

1 Richardson, *Material Ambitions*, esp. introduction.
2 Wendy Simonds, *Women and Self-Help Culture: Reading Between the Lines* (Rutgers University Press, 1992), 144.
3 Edward P. Thompson, *The Making of the English Working Class* (Penguin, 1991 [1963]), Ch. 11.
4 This is summed up persuasively by the concept of "emotionologies," a reference to each society's emotional standards and ideals. See Peter N. Stearns and Carol Z. Stearns, "Emotionology: Clarifying the History of Emotions and Emotional Standards," *American Historical Review*, 90 (1985).
5 Mohsen Joshanloo and Weijers Dan, "Aversion to Happiness across Cultures: A Review of Where and Why People are Averse to Happiness," *Journal of Happiness Studies*, 15 (2013).
6 Twenge and Campbell, *Narcissism Epidemic*, 17.

7   For criticism, see Cederström, *Happiness Fantasy*, 2.
8   Forbes, *Mindfulness and its Discontents*, 40.
9   Storr, *Selfie*, 328–9.
10 John P. Hewitt, *The Myth of Self-Esteem: Finding Happiness and Solving Problems in America* (St. Martin's Press, 1998), 142; Paul Dolan, *Happy Ever After: Escaping the Myth of the Perfect Life* (Allen Lane, 2019).
11 Oliver Burkeman, *The Antidote: Happiness for People Who Can't Stand Positive Thinking* (Canongate, 2012), 5.
12 Burkeman, *The Antidote*, 158–9.
13 "Self-Esteem," *Literary Speculum*, 2 (August 1822), 182.
14 "Self-Esteem," *Sixpenny Magazine*, 9 (April 1865), 472.
15 See Thomas Dixon, *From Passions to Emotions: The Creation of a Secular Psychological Category* (Cambridge University Press, 2003).
16 Silas Jones, *Practical Phrenology* (Russell, Shattuck, and Williams, 1836), 71.
17 "What is Self-Esteem?" *Saint Paul's Magazine*, 14 (March 1874), 336. Robert Owen was a Welsh textile manufacturer, philanthropist, and social reformer, and founder of utopian socialism and the cooperative movement.
18 See, among others, Roger Cooter, *Cultural Meaning of Popular Science: Phrenology and the Organization of Consent in Nineteenth-Century Britain* (Cambridge University Press, 2008); James Poskett, *Materials of the Mind: Phrenology, Race, and the Global History of Science, 1815–1920* (University of Chicago Press, 2019).
19 Franz Josef Gall, *On the Functions of the Brain and of Each of Its Parts or Organology*, trans. Winslow Lewis (Marsh, Capen, and Lyon, 1835), 167–70.
20 Hewett C. Watson, *Statistics of Phrenology* (Longman, Hurst, Rees, Orme, Brown & Green, 1836), 77–86.
21 Dixon, *From Passions to Emotions*.
22 Lunbeck, *Americanization of Narcissism*, 109–12.
23 See, among many others, Daniel J. Kevles, *In the Name of Eugenics: Genetics and the Uses of Human Heredity* (University of California Press, 1985); Daniel Pick, *Faces of Degeneration: Aspects of a European Disorder, c.1848–1918* (Cambridge University Press, 1989).
24 See, among others, Robert N. Proctor, *Racial Hygiene: Medicine under the Nazis* (Harvard University Press, 1988); Harry Bruinius, *Better for All the World: The Secret History of Forced Sterilization and America's Quest for Racial Purity* (Alfred A. Knopf, 2006). For emotions history perspectives, see chapters in Rob Boddice, *The Science of Sympathy: Morality, Evolution and Victorian Civilization* (University of Illinois Press, 2016).
25 For two contrasting perspectives, see Joel Whitebrook, *Freud: An Intellectual*

Biography (Cambridge University Press, 2017); Frederick Crews, *Freud: The Making of An Illusion* (Profile, 2017).

26 Sigmund Freud, *Civilization and its Discontents*, trans. Joan Riviere (Hogarth Press Institute of Psychoanalysis, 1930); see also Richard Overy, *The Morbid Age: Britain and the Crisis of Civilisation, 1919–1939* (Penguin, 2010), Ch. 4.

27 Cederström, *Happiness Fantasy*, 39–40.

28 Michael G. Johnson and Tracy B. Henley (eds.), *Reflections on the Principles of Psychology: William James after a Century* (Lawrence Erlbaum, 1990), 32.

29 For discussion, see, among others, Calum MacKellar and Christopher Bechtal (eds.), *The Ethics of the New Eugenics* (Berghahn Books, 2014); Judith Daar, *The New Eugenics: Selective Breeding in an Era of Reproductive Technologies* (Yale University Press, 2017).

30 Margaret J. Black and Stephen Mitchell, *Freud and Beyond: A History of Modern Psychoanalytic Thought* (Basic Books, 2016).

31 E.g. Warren Susman, "Personality and the Making of Twentieth-century Culture," in *Culture as History: The Transformation of American Society in the Twentieth Century* (Pantheon, 1984).

32 William James, *Reflections on the Principles of Psychology* (Henry Holt, 1890).

33 See Harrison, *Big Ego Trip*, 26.

34 Hewitt, *Myth of Self-Esteem*, 37.

35 Rose, *Inventing Our Selves*, 116–17, 75–6.

36 George H. Mead, *Mind, Self and Society* (University of Chicago Press, 1934).

37 See Hewitt, *Myth of Self-Esteem*, 39.

38 Edward Hoffman, *The Drive for Self: Alfred Adler and the Founding of Individual Psychology* (Addison-Wesley, 1994).

39 See, among other works, Alfred Adler, *The Neurotic Constitution*, trans. Bernard Glueck and John E. Lind (Moffat, Yard & Co., 1916); Alfred Adler, *The Practice and Theory of Individual Psychology*, trans. P. Radin (Kegan Paul, 1924).

40 Richard E. Watts, "Adler's Individual Psychology: The Original Positive Psychology," *Revista de Psicoterapia*, 26 (2015).

41 Alfred Adler, *The Science of Living* (George Allen and Unwin, 1930), 37.

42 Elizabeth Haiken, *Venus Envy: A History of Cosmetic Surgery* (Johns Hopkins University Press, 1997), 94, 111.

43 John H. Morgan, "The Interpersonal Psychotherapy of Harry Stack Sullivan: Remembering the Legacy," *Journal of Psychology and Psychotherapy*, 4 (2014).

44 See, among others, Henry Stack Sullivan, *The Interpersonal Theory of Psychiatry* (W.W. Norton, 1953).

45 Cushman, *Constructing the Self*, 159–209.

46 Karen Horney, *Our Inner Conflicts: A Constructive Theory of Neurosis* (W.W. Norton & Co., 1945); Karen Horney, *Neurosis and Human Growth: The Struggle Toward Self-Realization* (Routledge and Kegan Paul, 1950).
47 Jack L. Rubins, *Karen Horney: Gentle Rebel of Psychoanalysis* (Dial Press, 1978); Bernard J. Paris, *Karen Horney: A Psychoanalyst's Search for Self-Understanding* (Yale University Press, 1994), i–xix.
48 Rose, *Inventing Our Selves*, 22.
49 Jessica Grogan, *Encountering America: Humanistic Psychology, Sixties Culture, and the Shaping of the Modern Self* (Harper Perennial, 2013), 5–6.
50 Eva S. Moskowitz, *In Therapy We Trust: America's Obsession with Self Fulfillment* (Johns Hopkins University Press, 2001), 149.
51 Cushman, *Constructing the Self*, 245–6.
52 Dominick Cavallo, *A Fiction of the Past: The Sixties in American History* (St. Martin's Press, 1999), 58.
53 Wilkinson and Pickett, *Inner Level*.
54 "Beat Your Inferiority Complex," *Cavalcade* (December 1955), 6–7.
55 "Psychiatry and Beauty," *Cosmopolitan* (June 1956), 32.
56 Dale Carnegie, *How to Win Friends and Influence People* (Simon and Schuster, 1936); see also Barbara Ehrenreich, *Smile or Die: How Positive Thinking Fooled America and Changed the World* (Granta, 2009), 52.
57 Napoleon Hill, *Think and Grow Rich* (Ralston Books, 1937).
58 Melvin Powers, *Positive Thinking: The Technique for Achieving Self-Confidence and Success* (Wilshire, 1955), 5, 34, 52–3, 56.
59 Donald Meyer, *The Positive Thinkers: Religion as Pop Psychology from Mary Baker Eddy to Oral Roberts* (Pantheon, 1980 [1965]), 268.
60 Cederström and Spicer, *Wellness Syndrome*.
61 Peale, *Power of Positive Thinking*, 4–5.
62 Peale, *Power of Positive Thinking*, vii–ix, 1.
63 *St. Petersburg Times* (May 25, 1941). Adrienne L. McLean, *Being Rita Hayworth: Labor, Identity, and Hollywood Stardom* (Rutgers University Press, 2004), 105.
64 Quoted in Leo Verswijver, *Movies Were Always Magical* (McFarland, 2003), 180.
65 Steven Cohan, *Incongruous Entertainment: Camp, Cultural Value, and the MGM Musical* (Duke University Press, 2005), 297.
66 Quoted in Donald Spoto, *Marilyn Monroe: The Biography* (Rowman and Littlefield, 2001), 358.
67 See, among others, Steve Cohan, *Masked Men: Masculinity and the Movies in the Fifties* (Indiana University Press, 1997).
68 *The James Dean Story* (Warner Bros., 1957).

69 https://conorneill.com/2011/11/13/elvis-and-self-confidence/.
70 Quoted in Joel Williamson, *Elvis Presley: A Southern Life* (Oxford University Press, 2015), 197.
71 *Listen to Me Marlon* (Showtime Documentary Films, 2015).
72 Rose, *Inventing Our Selves*, 22.
73 Erich Fromm, *The Sane Society* (Holt, Rinehart, and Winston, 1955).
74 Betty Friedan, *The Feminine Mystique* (W.W. Norton & Co., 1963), esp. Ch. 5; Grogan, *Encountering America*, 260.
75 Georges Canguilhem, *On the Normal and the Pathological*, trans. Carolyn R. Fawcett (Reidel, 1978).
76 Harrison, *Big Ego Trip*, 41.
77 Adler, *Science of Living*, 39
78 Adler, *Science of Living*, 40.
79 Adler, *Science of Living*, 74–5.
80 Adler, *Science of Living*, 96–7.
81 Adler, *Science of Living*, 83.
82 Horney, *Neurosis and Human Growth*, 13, 17.
83 Cederström and Spicer, *Wellness Syndrome*.
84 Ehrenreich, *Smile or Die*, 8–9, 69, 89, 96.
85 Stephen Briers, *Psychobabble: Exploding the Myths of the Self-Help Generation* (Pearson, 2012), xi–xxiv.
86 Horney, *Neurosis and Human Growth*.
87 *Los Angeles Free Press* (December 5, 1969), 32.
88 Beatles, "Within You Without You" (Parlophone, 1967); Beatles, *Revolution* (Apple, 1968).
89 See, among others, Peter Doggett, *There's a Riot Going On: Revolutionaries, Rock Stars, and the Rise and Fall of '60s Counterculture* (Canongate, 2007).
90 Paul Oliver, *Hinduism and the 1960s: The Rise of a Counterculture* (Bloomsbury, 2014), 132, 178; Grogan, *Encountering America*, 238.
91 Quoted in Doggett, *There's a Riot Going On*, 56.
92 *Los Angeles Free Press* (June 18, 1965), 2.
93 Doggett, *There's a Riot Going On*, 102, 172.
94 Joachim C. Häberlen, *The Emotional Politics of the Alternative Left: West Germany, 1968–1984* (Cambridge University Press, 2018), 113.
95 Häberlen, *Emotional Politics*, 124–5, 170.
96 Moskowitz, *In Therapy We Trust*, 178–9.
97 Harrison, *Big Ego Trip*, 44–5; Grogan, *Encountering America*, 88.
98 Carl Rogers, *Client-Centered Therapy: Its Current Practice, Implications and Theory*, 3rd ed. (Houghton-Mifflin, 1956).

99 Biographies include David Cohen, *Carl Rogers: A Critical Biography* (Robinson, 2000); Carl R. Rogers and David E. Russell, *Carl Rogers: The Quiet Revolutionary, An Oral History* (Penmarin Books, 2002).
100 Nolan Jr., *Therapeutic State*, 3–4.
101 Carl Rogers, *On Becoming a Person* (Constable, 1967); Brian Thorne, *Carl Rogers* (SAGE, 1992), 24–31.
102 Abraham H. Maslow, *Toward a Psychology of Being*, 2nd ed. (Chapman and Hill, 1968), 5.
103 Edward Hoffman, *The Right to be Human: A Biography of Abraham Maslow* (Aquarian Press, 1989).
104 Abraham H. Maslow, "A Theory of Human Motivation," *Psychological Review*, 50 (1943); *Motivation and Personality* (Harper Row, 1954).
105 Herman, *Romance of American Psychology*, 270–1; Jeremy Carrette and Richard King, *Selling Spirituality: The Silent Takeover of Religion* (Routledge, 2005), 71, 76.
106 Ernest Becker, *The Birth and Death of Meaning: An Interdisciplinary Perspective on the Problem of Man*, 2nd ed. (Penguin, 1971 [1962]), 74, 75–6.
107 Nathaniel Branden, *The Psychology of Self-Esteem: A New Concept of Man's Psychological Nature* (Nash, 1969).
108 For detailed discussion, see, among others, Grogan, *Encountering America*.
109 Lasch, *Culture of Narcissism*, 9.
110 Ron Jacobs, *Daydream Sunset: The 60s Counterculture in the 70s* (CounterPunch, 2015), 91–2.
111 Doggett, *There's a Riot Going On*, 387.
112 Jacobs, *Daydream Sunset*, 2.
113 *Black Dwarf* (October 15, 1968), 8; (October 27, 1968), 6; (January 10, 1969), 3.
114 Doggett, *There's a Riot Going On*, 334–5.
115 John Lennon and Yoko Ono, *Some Time in New York City* (Apple, 1972).

Chapter 2: Race
1 E. Earl Baughman, Jr., *Black Americans: A Psychological Analysis* (Academic Press, 1971), 46.
2 Wright, *Rise of the Therapeutic Society*.
3 Alondra Nelson, *Body and Soul: The Black Panther Party and the Fight against Medical Discrimination* (University of Minnesota, 2011).
4 Lisa M. Corrigan, *Black Feelings: Race and Affect in the Long Sixties* (University Press of Mississippi, 2020).
5 Daryl Michael Scott, *Contempt and Pity: Social Policy and the Image of the*

Damaged Black Psyche, 1880–1996 (University of North Carolina Press, 1997), 81–6.
6 Gabriel Mendes, "Race," in Aubry and Travis, *Rethinking Therapeutic Culture*, 61.
7 Eddie S. Glaude Jr., *Democracy in Black: How Race Still Enslaves the American Soul* (Crown, 2017), 18.
8 Robin D.G. Kelley, *Freedom Dreams: The Black Radical Imagination* (Beacon Press, 2002).
9 Isabel Wilkerson, *Caste: The Origins of Our Discontents* (Random House, 2020), 141.
10 Studies of race and psychology include Graham Richards, *"Race," Racism and Psychology: Towards A Reflexive History* (Routledge, 1997); Scott, *Contempt and Pity*; Jay Garcia, *Psychology Comes to Harlem: Rethinking the Race Question in Twentieth-Century America* (Johns Hopkins University Press, 2012). See also Rana Hogarth, *Medicalizing Blackness: Making Racial Difference in the Atlantic World, 1780–1840* (University of North Carolina Press, 2017).
11 Angela Saini, *Inferior: The True Power of Women and the Science that Shows It* (4th Estate, 2017).
12 Caroline Criado Perez, *Invisible Women: Exposing Data Bias in a World Designed for Men* (Vintage, 2019).
13 I borrow here from Scott, *Contempt and Pity*.
14 Texts include Nancy Stepan, *The Idea of Race in Science: Great Britain, 1800–1960* (Palgrave Macmillan, 1982); Kenan Malik, *The Meaning of Race: Race, History, and Culture in Western Society* (New York University Press, 1996); Warwick Anderson, *The Cultivation of Whiteness: Science, Health and Racial Disparity in Australia* (Basic Books, 2003).
15 W.E.B. Du Bois, *The Souls of Black Folk* (Bantam, 1989 [1903]), 142.
16 hooks, *Rock My Soul*, 3.
17 Gavin Schaffer, *Racial Science and British Society, 1930–62* (Palgrave Macmillan, 2008); Marek Kohn, *The Race Gallery: The Return of Racial Science* (Vintage, 1996).
18 Charles H. Wesley, "The Concept of Negro Inferiority in American Thought," *Journal of Negro History*, 4 (1940), 560.
19 James W. Vander Zanden, "The Ideology of White Supremacy," *Journal of the History of Ideas*, 30 (1959).
20 E. Franklin Frazier, *The Negro in the United States*, 2nd ed. (Toronto: Macmillan Company, 1957 [1949]); Richards, *"Race," Racism and Psychology*, 71–2; Garcia, *Psychology Comes to Harlem*, 9; Eddie S. Glaude, Jr., *Begin Again: James Baldwin's America and its Urgent Lessons for Today* (Crown, 2020), xxv.

21 E.g. Oliver C. Cox, *Caste, Class and Race* (Doubleday, 1948), 382–3.
22 Richards, *"Race," Racism and Psychology*, 151, 242.
23 Quoted in William E. Cross Jr., *Shades of Black: Diversity in African-American Identity* (Temple University Press, 1991), ix.
24 Richards, *"Race," Racism and Psychology*, 242.
25 Abram Kardiner and Lionel Ovesey, *The Mark of Oppression: A Psychosocial Study of the American Negro* (W.W. Norton & Co., 1951), 301.
26 Everett V. Stonequist, *The Marginal Man: A Study in Personality and Culture Conflict* (Charles Scribner's Sons, 1937), 139–40.
27 Kardiner and Ovesey, *Mark of Oppression*, 310.
28 James E. Conyers and T.H. Kennedy, "Negro Passing: To Pass or Not to Pass," *Phylon: The Atlanta University Review of Race and Culture*, 24 (1963). For analysis, see Gayle Ward, *Crossing the Line: Racial Passing in Twentieth-Century US Literature and Culture* (Duke University Press, 2000).
29 Stonequist, *The Marginal Man*, 184.
30 Harold Proshansky and Peggy Newton, "The Nature and Meaning of Negro Self-Identity," in Martin Deutsch, Irwin Katz, and Arthur R. Jensen (eds.), *Social Class, Race and Psychological Development* (Holt, Rinehart, and Winston, 1968), 195.
31 See Richards, *"Race," Racism and Psychology*, 139–40.
32 Scott, *Contempt and Pity*, 76–9.
33 Thomas F. Pettigrew, *A Profile of the Negro American* (D. Van Nostrand, 1964), 9.
34 Robert Coles, *Children of Crisis: A Study of Courage and Fear* (Little, Brown and Co., 1967).
35 Richards, *"Race," Racism and Psychology*, 241.
36 Scott, *Contempt and Pity*, 93–4, 107–8.
37 George Yancy, *Black Bodies, White Gaze: The Continuing Significance of Race in America*, 2nd ed. (Rowman and Littlefield, 2017 [2009]).
38 Coles, *Children of Crisis*.
39 Kenneth B. Clark, *Effect of Prejudice and Discrimination on Personality Development: Confidential Draft for the Use of the Technical Community on Fact Finding* (Schoen Books, 1950)
40 Kenneth B. Clark and Mamie K. Clark, "Racial Identification and Preference in Negro Children," in Theodore M. Newcomb and Eugene L. Hartley (eds.), *Readings in Social Psychology* (Holt, Rinehart, and Winston, 1947); Kenneth B. Clark, "How Children Learn about Race," reprinted in Clayborne Carson, David J. Garrow, Gerald Gill, et al., *The Eyes on the Prize: Civil Rights Reader* (Penguin, 1991).

41 Gwen Bergner, "Black Children, White Preference: *Brown v. Board*, the Doll Tests and the Politics of Self-Esteem," *American Quarterly*, 61 (2009).
42 *Brown v. Board of Education* of Topeka, 347 U.S. 483 (1954).
43 Quoted in Garcia, *Psychology Comes to Harlem*, 7.
44 Herman, *Romance of American Psychology*, 197, 199; Scott, *Contempt and Pity*, 138.
45 Cross Jr., *Shades of Black*, 35–6.
46 Derrick Bell, *Silent Covenants: Brown v. Board of Education and the Unfulfilled Hopes for Racial Reform* (Oxford University Press, 2004).
47 Kenneth B. Clark, *Dark Ghetto: Dilemmas of Social Power* (Victor Gollancz, 1965), 65–7.
48 Clark, *Dark Ghetto*, 63–4.
49 Bergner, "Black Children, White Preference," 308.
50 Julia A. Boyd, *In the Company of My Sisters: Black Women and Self-Esteem* (Dutton, 1993), x–xi.
51 Boyd, *In the Company of My Sisters*, 1–2.
52 Ibram X. Kendi, *Stamped from the Beginning: The Definitive History of Racist Ideas in America* (Bodley Head, 2016), 394.
53 Frantz Fanon, *Black Skin, White Masks* (Grove Press, 1952), 69, 75, 82–3.
54 Arnold M. Rose, *The Negro's Morale* (University of Minnesota Press, 1949), 55, 5.
55 George Eaton Simpson and J. Milton Yinger, *Racial and Cultural Minorities*, 3rd ed. (Harper and Row, 1953), 153.
56 Adam Fairclough, *Better Day Coming: Blacks and Equality, 1890–2000* (Viking, 2001).
57 Rose, *Negro's Morale*, 95.
58 Quoted in Fairclough, *Better Day Coming*, 226.
59 August Meier, Elliott Rudwick, and Francis L. Broderick, *Black Protest Thought in the Twentieth Century*, 2nd ed. (Bobbs-Merrill Educational Publishing, 1971 [1965]), 291, 293–4.
60 Jeanne Theoharis, "'A Life History of Being Rebellious': The Radicalism of Rosa Parks," in Dayo F. Gore, Jeanne Theoharis, and Komozi Woodard (eds.), *Want to Start a Revolution? Radical Women in the Black Freedom Struggle* (New York University Press, 2009), 116.
61 Jacqueline Jones, *Labor of Love, Labor of Sorrow: Black Women, Work, and the Family from Slavery to the Present* (Basic Books, 1985), 279–80.
62 Michelle Alexander, *The New Jim Crow: Mass Incarceration in the Age of Color-Blindness* (Penguin, 2019 [2010]), 282.
63 Harold Proshansky and Peggy Newton, "The Nature and Meaning of Negro Self-Identity," in Deutsch et al. (eds.), *Social Class*, 225.

64 Kardiner and Ovesey, *Mark of Oppression*, 304, 308.
65 Scott, *Contempt and Pity*, 99.
66 Fairclough, *Better Day Coming*, 307.
67 Malcolm X, *The Autobiography of Malcolm X* (London: Penguin, 1965), 81, 82–3.
68 Quoted in Robert L. Scott and Wayne Brockreide, *The Rhetoric of Black Power* (Harper and Row, 1969), 43–4.
69 Kendi, *How to Be an Anti-Racist*, 28.
70 Kendi, *Stamped from the Beginning*, 395–7.
71 E. Franklin Frazier, *Black Bourgeoisie* (The Free Press, 1957); Stonequist, *Marginal Man*.
72 Herman, *Romance of American Psychology*, 199.
73 Stokely Carmichael and Charles V. Hamilton, *Black Power: The Politics of Liberation in America* (Penguin, 1971 [1967]), 46–8.
74 Baughman, *Black Americans*, 42.
75 Alvin F. Poussaint, *Why Blacks Kill Blacks* (Emerson Hall, 1972), 26.
76 Carmichael and Hamilton, *Black Power*, 69.
77 Carmichael and Hamilton, *Black Power*, 93.
78 Jeffrey O.G. Ogbar, *Black Power: Radical Politics and African American Identity* (Johns Hopkins University Press, 2004), 93.
79 Poussaint, *Why Blacks Kill Blacks*, xi–xii.
80 Alvin F. Poussaint, "The Negro American: His Self-Image and Integration," in Floyd B. Barbour (ed.), *The Black Power Revolt: A Collection of Essays* (Collier-Macmillan, 1968), 106–9.
81 Poussaint, "The Negro American," 111–15; *Why Blacks Kill Blacks*, 27–8.
82 John O'Killens, "Explanation of the Black Psyche," *New York Times* (June 7, 1964).
83 Nelson, *Body and Soul*.
84 Quoted in Benjamin Muse, *The American Negro Revolution: From Nonviolence to Black Power, 1963–1967* (Indiana University Press, 1968), 199.
85 Robin D.G. Kelley, *Race Rebels: Culture, Politics, and the Black Working Class* (Free Press, 1994), 56–7.
86 O'Killens, "Explanation of the Black Psyche."
87 Kathy Peiss, *Hope in a Jar: The Making of America's Beauty Culture* (Henry Holt, 1998), 258.
88 *Autobiography of Malcolm X*, 139.
89 Ta-Nehisi Coates, *We Were Eight Years in Power: An American Tragedy* (Penguin, 2017), 94.
90 Imani Perry, *Prophets of the Hood: Politics and Poetics in Hip Hop* (Duke University Press, 2004), 4.

91 LeRoi Jones, *Blues People: The Negro Experience in White America and the Music that Developed from It* (William Morrow, 1963), 142; James A. Williams, *Funk Music: A Critical Enquiry* (Createspace, 2015), 84.
92 Charles Keil, *Urban Blues* (University of Chicago Press, 1966).
93 Frank Kofsky, *Black Nationalism and the Revolution in Music* (Pathfinder, 1970), 9.
94 Ogbar, *Black Power*, 112.
95 Williams, *Funk Music*, 34–5.
96 James Brown, *Black and Proud* (King Records, 1968).
97 Glaude, *Begin Again*, 94–5.
98 Novotoy Lawrence, Jr. and Gerald R. Butters (eds.), *Beyond Blaxploitation* (Wayne State University Press, 2016).
99 Eldridge Cleaver, *Soul on Ice* (Delta, 1968), 92–3.
100 Eddie S. Glaude Jr. (ed.), *Is It Nation Time? Contemporary Essays on Black Power and Black Nationalism* (University of Chicago Press, 2002).
101 Jones, *Labor of Love*, 311.
102 Keisha N. Blain, *Set the World on Fire: Black Nationalist Women and the Global Struggle for Freedom* (University of Pennsylvania Press, 2018), 8; Jones, *Labor of Love*, 279–80.
103 Cleaver, *Soul on Ice*, 12, 26–7.
104 Cleaver, *Soul on Ice*, 97–100.
105 hooks, *Rock My Soul*, 7.
106 "Riots Seen as Bid for Self-Esteem," *New York Times* (July 30, 1967), 49.
107 Baughman, *Black Americans*, 38.
108 Keil, *Urban Blues*, 13.
109 Morris Rosenberg and Robert G. Simmons, *Black and White Self-Esteem: The Urban School Child* (American Sociological Association, 1971).
110 I draw here from Barry D. Adam, "Inferiorization and Self-Esteem," *Social Psychology*, 41 (1978).
111 Baughman, *Black Americans*, 40; Susan Harris and John Braun, "Self-Esteem and Racial Preference in Black Children," *American Journal of Orthopsychiatry*, 42 (1972).
112 Quoted in Poussaint, *Why Blacks Kill Blacks*, 25–6.
113 hooks, *Rock My Soul*, xii, 161.
114 Glaude Jr., *Democracy in Black*, 30.
115 Kendi, *Stamped from the Beginning*, 8.
116 Andrew Hacker, *Two Nations: Black and White, Separate, Hostile, Unequal* (Charles Scribner's Sons, 1992), 19.
117 Kendi, *Stamped from the Beginning*, 429.

118 Alexander, *The New Jim Crow*; hooks, *Rock My Soul*, 72.
119 Alexander, *The New Jim Crow*, 206–15.
120 Mary-Frances Winters, *Black Fatigue: How Racism Erodes the Mind, Body and Spirit* (Berrett-Koehler, 2020), 2.
121 hooks, *Rock My Soul*, xii–xiii.
122 hooks, *Rock My Soul*, x.

Chapter 3: Sexuality

1 John D'Emilio, *Sexual Politics, Sexual Communities: The Making of a Homosexual Minority in the United States, 1940–70* (University of Chicago Press, 1983), 23, 39.
2 See, among others, Martin Duberman, *Stonewall: The Definitive Story of the LGBTQ Rights Uprising that Changed America* (Plume, 2019); Matthew Todd, *Pride: The Story of the LTBTQ Equality Movement* (Welbeck, 2019).
3 Ilan H. Meyer, "Minority Stress and Mental Health in Gay Men," *Journal of Health and Social Behavior*, 36 (1995).
4 Deborah Gould, "Rock the Boat, Don't Rock the Boat, Baby: Ambivalence and the Emergence of Militant AIDS Activism," in Jeff Goodwin, James M. Jasper, and Francesca Polletta (eds.), *Passionate Politics: Emotions and Social Movements* (University of Chicago Press, 2001), 139. See also Heather Love, *Feeling Backward: Loss and the Politics of Queer History* (Harvard University Press, 2007), 1–3; David Halperin and Valerie Traub (eds.), *Gay Shame* (University of Chicago Press, 2010).
5 Craig Griffiths, *The Ambivalence of Gay Liberation: Male Homosexual Politics in 1970s West Germany* (Oxford University Press, 2021), 2, 9, 18.
6 Merle Dale Miller, *On Being Different: What It Means to be Homosexual* (Random House, 1971), 39.
7 Griffiths, *Ambivalence of Gay Liberation*, 197.
8 Odets, *In the Shadow of the Epidemic*, 127.
9 I borrow here from Margot Canaday, *The Straight State: Sexuality and Citizenship in Twentieth-Century America* (Princeton University Press, 2009), 1–2.
10 Lisa McGirr, *Suburban Warriors: The Origins of the New American Right* (Princeton University Press, 2001).
11 Nan Alamilla Boyd, *Wide Open Town: A History of Queer San Francisco to 1965* (University of California Press, 2003), 1–2.
12 Edmund Bergler, *Homosexuality: Disease or Way of Life?* (Collier Books, 1956), 27.
13 Roger Blake, *The Making of a Homosexual* (Century Books, 1967), 15, 18.

14 Henry L. Minton, *Departing from Deviance: A History of Homosexual Rights and Emancipatory Science in America* (University of Chicago Press, 2001).
15 Charles W. Socarides, *The Overt Homosexual* (Grune and Stratton, 1968), 7.
16 Allan Bérubé, *Coming Out Under Fire: The History of Gay Men and Women in World War Two*, 2nd ed. (University of North Carolina Press, 2010 [1990]), 2.
17 For discussion, see Stephanie Coontz, *The Way We Never Were: American Families and the Nostalgia Trap* (Basic Books, 1993).
18 David K. Johnson, *The Lavender Scare: The Cold War Persecution of Gays and Lesbians in the Federal Government* (University of Chicago Press, 2009). See also Tommy Dickinson, *"Curing Queers": Mental Nurses and their Patients, 1935–74* (Manchester University Press, 2015), Ch. 1.
19 Anna Lvosky, *Vice Patrol: Cops, Courts, and the Struggle over Gay Life before Stonewall* (University of Chicago Press, 2021), 31.
20 Robert Allan Harper, "Psychological Aspects of Homosexuality," *Advances in Sex Research*, 1 (1963), 188.
21 Robert Allan Harper, "Can Homosexuals Be Changed?" in Isadore Rubin (ed.), *Homosexuals Today* (Heath, 1965), 69.
22 Harper, "Psychological Aspects of Homosexuality," 192–4.
23 Harper, "Can Homosexuals Be Changed?" 72.
24 Harper, "Psychological Aspects of Homosexuality," 195–6.
25 Harper, "Psychological Aspects of Homosexuality," 196–7.
26 See, among others, Dickinson, *"Curing Queers"*; Kate Davison, "Cold War Pavlov: Homosexual Aversion Therapy in the 1960s," *History of the Human Sciences*, 34 (2021).
27 Griffiths, *Ambivalence of Gay Liberation*, 6.
28 Julian Jackson, "The Homophile Movement," in David Paternotte and Manon Tremblay (eds.), *The Ashgate Companion to Lesbian and Gay Activism* (Routledge, 2015), 31.
29 Boyd, *Wide Open Town*, esp. Ch. 4.
30 Laura A. Belmonte, *The International LGBT Rights Movement: A History* (Bloomsbury, 2021), 105.
31 D'Emilio, *Sexual Politics, Sexual Communities*, 58–61, 64.
32 Donald Webster Cory, *The Homosexual in America: A Subjective Approach* (Greenberg, 1951).
33 Donald Webster Cory, *The Homosexual Outlook* (Henderson and Spalding, 1953), xiii.
34 Cory, *The Homosexual Outlook*, 11.

35 Cory, *The Homosexual Outlook*, 10, 13.
36 Wainwright Churchill, *Homosexual Behavior among Males: A Cross-Cultural and Cross-Species Investigation* (Hawthorn Books, 1967).
37 Miriam G. Reumann, *American Sexual Character: Sex, Gender, and National Identity in the Kinsey Reports* (University of California Press, 2005), 1.
38 Reumann, *American Sexual Character*, 165–7.
39 "A Matter of Self-Esteem," *Mattachine*, 9 (1963), 11.
40 Donald J. West, *Homosexuality* (Penguin, 1960 [1955]), 11.
41 Martin Meeker, *Contacts Desired: Gay and Lesbian Communications and Community, 1940s–1970s* (University of Chicago Press, 2006).
42 Reumann, *American Sexual Character*, 171–2.
43 *Vector*, 1 (December 1964), 1.
44 "On Getting and Using Power," *Vector*, 1 (January 1965), 4.
45 Jackson, "The Homophile Movement," 34.
46 Belmonte, *International LGBT Rights Movement*, 121.
47 Jackson, "The Homophile Movement," 34.
48 "Marie Angell Addresses SIR," *Vector*, 1 (June 1965), 1, 8.
49 Frank J. Howell, "Toward a Positive View of the Functioning Homosexual," *Vector*, 2 (March 1966), 11.
50 "Self-Respect Required," *Vector*, 2 (July 1966), 4.
51 D'Emilio, *Sexual Politics, Sexual Communities*.
52 Lvosky, *Vice Patrol*, Ch. 6.
53 "That Time/Life Look," *Vector*, 3 (February 1967).
54 *Vector*, 3 (June 1967), 23.
55 Miller, *On Being Different*, 33.
56 See, among others, Barry D. Adam, *The Rise of a Lesbian and Gay Movement* (Twayne, 1995).
57 "God Save the Queen," *Gay Sunshine*, 1 (November 1970), 2.
58 "Our Bodies Say Yes, Society Says No," *Gay Sunshine*, 1 (October 1970), 11.
59 *Tangents*, 3 (October 1968), 23.
60 *Tangents*, 4 (October 1969), 28.
61 Allen Young, "Out of the Closets, Into the Streets," in Karla Jay and Allen Young (eds.), *Out of the Closets: Voices of Gay Liberation* (Jove/HBJ, 1977 [1972]), 19.
62 Bergler, *Homosexuality*; Irving Bieber, *Homosexuality: A Psychoanalytic Study* (Basic Books, 1962); Albert Ellis, *Homosexuality: Its Causes and Cures* (Lyle Stuart Inc., 1965); Socarides, *The Overt Homosexual*; Lionel Ovesey, *Homosexuality and Pseudo-Homosexuality* (Science House, 1969); Lawrence J. Hatterer, *Changing Homosexuality in the Male* (Delta, 1971).

63 Leo Louis Martello, "The One Thing Every Human Needs is a Sense of Self-Esteem," *Gay* (April 13, 1970), 15.
64 "Feeling Good," *Update* (May 4, 1979), 17.
65 Jerold S. Greenberg, "The Effects of a Homophile Organization on the Self-Esteem and Alienation of its Members," *Journal of Homosexuality*, 1 (1976), 313–17.
66 D'Emilio, *Sexual Politics, Sexual Communities*, 100.
67 Jackson, "The Homophile Movement," 33.
68 *Lesbian Connection*, 2 (July 1976), 3–4.
69 *Amazon Quarterly*, 1 & 2 (Fall 1972), 10, 6.
70 *Amazon Quarterly*, 1 & 2 (Fall 1972), 55.
71 D'Emilio, *Sexual Politics, Sexual Communities*, 101.
72 Del Martin and Phyllis Lyon, *Lesbian/Woman* (Bantam, 1972), 27.
73 D'Emilio, *Sexual Politics, Sexual Communities*, 1; see also Joel D. Hencken, "Homosexuality and Psychoanalysis: Toward a Mutual Understanding," in William Paul, James D. Weinrich, John C. Gonsiorek, and Mary E. Hotvedt (eds.), *Homosexuality: Social, Psychological and Biological Issues* (SAGE, 1982), 138–9.
74 Edwin M. Schur, *Crimes without Victims: Deviant Behavior and Public Policy* (Prentice Hall, 1965), 95–102.
75 Eric Brandt (ed.), *Dangerous Liaisons: Blacks, Gays, and the Struggle for Equality* (New Press, 1999), 2.
76 Henry Louis Gates Jr., "Foreword," in Delroy Constantine-Simms (ed.), *The Greatest Taboo: Homosexuality in Black Communities* (Alyson Books, 2000), xi.
77 Miller, *On Being Different*, 38.
78 Cleaver, *Soul on Ice*, 106, 100.
79 Kevin J. Mumford, *Not Straight, Not White: Black Gay Men from the March on Washington to the AIDS Crisis* (University of North Carolina Press, 2016), 5.
80 Earl Ofari Hutchinson, "My Gay Problem, Your Black Problem," in Constantine-Simms, *Greatest Taboo*, 2–3.
81 D'Emilio, *Sexual Politics, Sexual Communities*, 192.
82 "Gay is Good," *Vector*, 4 (November 1968), 5.
83 "Gay Voices from the Third World: Are We Listening?" *Advocate* (January 24, 1984), 26–9.
84 See, among others, Ronald Bayer, *Homosexuality and American Psychiatry: The Politics of Diagnosis* (Princeton University Press, 1987); Jennifer Terry, *An American Obsession: Science, Medicine, and the Place of Homosexuality in Modern Society* (University of Chicago Press, 1999).
85 Clarence A. Tripp, *The Homosexual Matrix* (Quartet, 1977 [1975]).

86 "There Are No Cures. Aversion Therapy Pioneer Takes a Second Look," *Advocate*, 191 (June 2, 1976), 8–9.
87 Charles Silverstein, "Psychological and Medical Treatments of Homosexuality," in John C. Gonsiorek and James D. Weinrich (eds.), *Homosexuality: Research Implications for Public Policy* (SAGE, 1991), 111.
88 Peter Hegarty, *A Recent History of Lesbian and Gay Psychology* (Routledge, 2018), 15.
89 Betty Berzon and Robert Leighton (eds.), *Positively Gay* (Celestial Arts, 1979).
90 "Therapy Must Be Positive, Not Negative," *Advocate*, 133 (March 13, 1974), 37–8.
91 Rob Eichberg, "Hearing the Distant Drums," *Advocate* (September 15, 1987), 9, 20–1.
92 Barry D. Adam, "Inferiorization and Self-Esteem," *Social Psychology*, 41 (1978), 49.
93 Evelyn Hooker, "The Adjustment of the Male Overt Homosexual," *Journal of Projective Techniques*, 21 (1956), 18–31.
94 William E. Cross Jr., "The Negro-to-Black Conversion Experience: Toward a Psychology of Black Liberation," *Black World*, 20 (1971); Thomas A. Parham and Janet E. Helms, "The Influence of Black Students' Racial Identity Attitudes on Preference for Counselor's Race," *Journal of Counseling Psychology*, 28 (1981). Karina L. Walters and Jane M. Simoni, "Lesbian and Gay Male Group Identity Attitudes and Self-Esteem: Implications for Counseling," *Journal of Counseling Psychology*, 40 (1993). A. Elfin Moses and Robert O. Hawkins, Jr., *Counseling Lesbian Women and Gay Men: A Life-Issues Approach* (C.V. Mosby Company, 1982).
95 Don Clark, *Loving Someone Gay* (New American Library, 1977), 19, 22, 26–7. For a recent autobiography, see Don Clark, *Someone Gay: Memoirs* (GayTrueWords Press, 2020).
96 Clark, *Loving Someone Gay*, 29, 31.
97 Clark, *Loving Someone Gay*, 37, 39.
98 Clark, *Loving Someone Gay*, 41–2, 1.
99 Richard A. Isay, *Becoming Gay: The Journey to Self-Acceptance* (Pantheon Books, 1996), 7.
100 Isay, *Becoming Gay*, 11, 30–1.
101 E.g. John P. De Cecco, *Bashers, Baiters and Bigots: Homophobia in American Society* (Harrington Park Press, 1985).
102 George Weinberg, *Society and the Healthy Homosexual* (St. Martin's Press, 1972), 2.
103 Gregory M. Herek, "Beyond Homophobia: A Social Psychological Perspective

of Attitudes towards Lesbians and Gay Men," *Journal of Homosexuality*, 1–2 (1984).

104 E.g. Eli Coleman, "Developmental Stages of the Coming-Out Process," in Paul et al., *Homosexuality*, 150.

105 Ritch C. Savin-Williams, "Parental Influences on the Self-Esteem of Gay and Lesbian Youths: A Reflected Appraisals Model," *Journal of Homosexuality*, 17 (1989).

106 Natalie Jane Woodman and Harry R. Lenna, *Counseling with Gay Men and Women* (Jossey-Bass, 1980), 46; Nanette Gartrell, "Combating Homophobia in the Psychotherapy of Lesbians," *Women and Therapy*, 3 (1984).

107 Karen M. Jordan and Robert H. Deluty, "Coming Out for Lesbian Women: Its Relation to Anxiety, Positive Affectivity, Self-Esteem and Social Support," *Journal of Homosexuality*, 35 (1998), 41; Joel W. Wells and William B. Kline, "Self-disclosure of Homosexual Orientation," *Journal of Social Psychology*, 127 (1987).

108 Canaday, *The Straight State*.

109 Vivienne C. Cass, "Homosexual Identity Formation: Testing a Theoretical Model," *Journal of Sex Research*, 20 (1984).

110 E.g. John Vincke and Ralph Bolton, "Social Support, Depression and Self-Acceptance among Gay Men," *Human Relations*, 47 (1994).

111 See, among others, Mirko D. Grmek, *History of AIDS: Emergence and Origin of a Modern Pandemic* (Princeton University Press, 1990); Odets, *In the Shadow of the Epidemic*, 7–8; Dean L. Wolcott, Sheila Namir, Fawzy I. Fawzy, et al., "Illness Concerns, Attitudes towards Homosexuality and Social Support in Gay Men with AIDS," *General Hospital Psychiatry*, 8 (1986); Deborah B. Gould, *Moving Politics: Emotion and ACT UP's Fight Against AIDS* (University of Chicago Press, 2009), 10, 63; Steven Epstein, *Impure Science: AIDS, Activism, and the Politics of Knowledge* (University of California Press, 1998).

112 Virginia Lehman and Noreen Russell, "Psychological and Social Issues of AIDS and Strategies for Survival," in Victor Gong (ed.), *Understanding AIDS: A Comprehensive Guide* (Rutgers, 1985), 178.

113 Cited in Stephen F. Morin, Kenneth A. Charles, and Alan K. Malyon, "The Psychological Impact of AIDS on Gay Men," *American Psychologist*, 39 (1984), 1291.

114 For the broader picture, see David France, *How to Survive a Plague: The Story of How Activists and Scientists Tamed AIDS* (Picador, 2017).

115 Patti Lather and Chris Smithies, *Troubling the Angels: Women Living with HIV/AIDS* (Westview Press, 1997), 5.

116 Mervyn F. Silverman, "Introduction: What We Have Learned," in Leon

McKusick (ed.), *What to Do about AIDS: Physicians and Mental Health Professionals Discuss the Issues* (University of California Press, 1986), 1–9.
117 Odets, *In the Shadow of the Epidemic*, 9, 11–19.
118 Odets, *In the Shadow of the Epidemic*, 126, 160–2.
119 Geraldo Lima, Charles T. Lo Presto, Martin F. Sherman, and Steven A. Sobelman, "The Relationship Between Homophobia and Self-Esteem in Gay Males with AIDS," *Journal of Homosexuality*, 25 (1993), 74.
120 Odets, *In the Shadow of the Epidemic*, 230–1.
121 Norris G. Lang, "Stigma, Self-Esteem, and Depression: Psychosocial Responses to Risk of AIDS," *Human Organization*, 50 (1991).
122 Odets, *In the Shadow of the Epidemic*, 104–5.
123 "Study Links Homophobia, Self-Esteem, and HIV Risk among African Americans," *AIDS Weekly Plus* (10 August 1998), 46.
124 Cathy J. Cohen, *The Boundaries of Blackness: AIDS and the Breakdown of Black Politics* (University of Chicago Press, 1999), ix–xi, 23, 35, 50, 94, 186.
125 "TV AIDS Commercial Features Gay Self-Esteem while Encouraging Safer Sex," *Wisconsin Light* (March 30, 1995), 10.
126 "New San Francisco AIDS Prevention Hope: Raise Self-Esteem," *Out Now!* (April 4, 1995), 1, 4.
127 "The Glamorization of AIDS," *Advocate*, 695 (November 28, 1995).
128 Alan Downs, *The Velvet Rage: Overcoming the Pain of Growing Up Gay in a Straight Man's World*, 2nd ed. (Da Capo, 2012 [2005]), xi, xii.
129 Odets, *Out of the Shadows*, 60, 84, 68, 73.

Chapter 4: Mothers
1 Anderson J. Franklin and Nancy Boyd-Franklin, "A Psychoeducational Perspective on Black Parenting," in Harriette Pipes McAdoo and John Lewis McAdoo (eds.), *Black Children: Social, Educational and Parental Environments* (SAGE, 1985), 205.
2 Jacobs, *Daydream Sunset*; Joseph Heath and Andrew Potter, *The Rebel Sell: How the Counterculture became Consumer Culture* (Capstone, 2005).
3 I borrow here from Cederström, *Happiness Fantasy*, 14, 64.
4 See, among others, John Ehrman, *The Rise of Neoconservatism: Intellectuals and Foreign Affairs, 1945–1994* (Yale University Press, 1995); Gregory L. Schneider, *Cadres for Conservatism: Young Americans for Freedom and the Rise of the Contemporary Right* (New York University Press, 1999); McGirr, *Suburban Warriors*.
5 Phyllis Schlafly, *Feminist Fantasies* (Spence Pub, 2003); *Who Killed the American Family?* (WND Books, 2014).

6 Jeanne Theoharis and Athan Theoharis, *These Yet to Be United States: Civil Rights and Civil Liberties in America since 1945* (Wadsworth, 2003), 170–1.
7 Theoharis and Theoharis, *These Yet to Be United States*, 124–5; Keeanga-Yamahtta Taylor, *From #Blacklivesmatter to Black Liberation* (Haymarket Books, 2016), 70; Theoharis and Theoharis, *These Yet to Be United States*, 170–1, 192.
8 Robin Diangelo, *White Fragility: Why It's So Hard for White People to Talk about Racism* (Allen Lane, 2019), 34, 27–8.
9 Diangelo, *White Fragility*, 10.
10 Kenzo E. Bergeron, *Challenging the Cult of Self-Esteem in Education: Education, Psychology, and the Subaltern Self* (Routledge, 2018), 9, 49.
11 Imani Perry, *More Beautiful and More Terrible: The Embrace and Transcendence of Racial Inequality in the United States* (New York University Press, 2011).
12 Peter N. Stearns, *Anxious Parents: A History of Modern Childrearing in America* (New York University Press, 2003), 1–4.
13 Viviana Zelizer, *Pricing the Priceless Child: The Changing Social Value of Children* (Princeton University Press, 1985).
14 Coontz, *The Way We Never Were*, 26.
15 Jean Illsley Clarke, *Self-Esteem: A Family Affair* (Winston Press, 1978), vii.
16 Elizabeth Badinter, *L'Amour en plus: Histoire de l'amour maternel* (Falmmarian, 1980).
17 Michael E. Staub, *Madness is Civilization: When the Diagnosis was Social, 1948–1980* (University of Chicago Press, 2011), 66.
18 Phyllis Schlafly, *The Power of the Positive Woman* (Crown Publishing, 1977); Brigitte Berger, *The Family in the Modern Age: More than a Lifestyle Choice* (Transaction, 2002), 19–30; Elaine Tyler May, *Homeward Bound: American Families in the Cold War Era* (Basic Books, 2008 [1988]), 9, 17, 29–30.
19 Ruth Feldstein, *Motherhood in Black and White: Race and Sex in American Liberalism, 1930–1965* (Cornell University Press, 2000), 169; Staub, *Madness is Civilization*, 140–1.
20 Furedi, *Therapy Culture*, 66.
21 Deborah Weinstein, *The Pathological Family: Postwar America and the Rise of Family Therapy* (Cornell University Press, 2013), 5–6.
22 Coontz, *The Way We Never Were*, 27, 32, 97.
23 Teri Chettiar, "Democratizing Mental Health: Motherhood, Therapeutic Community and the Emergence of the Psychiatric Family at the Cassel Hospital in Post-Second World War Britain," *History of the Human Sciences*, 25 (2012), 108.
24 Staub, *Madness is Civilization*, 23; see also Staub's introduction and pp. 23–5.

25 Berger, *Family in the Modern Age*, 7–8.
26 David Elkind, "'Good Me' or 'Bad Me': The Sullivan Approach to Personality," *New York Times* (September 24, 1972), SM18.
27 *Developing Self-Esteem* (Ross Laboratories, 1969), 4.
28 Erik Erikson, *Childhood and Society* (W.W. Norton & Co., 1950).
29 May, *Homeward Bound*, 9, 17, 29–30; Christina Hardyment, *Perfect Parents: Baby-Care Advice Past and Present* (Oxford University Press, 1995), 223–7.
30 John Bowlby, *Maternal Care and Mental Health* (Schocken Books, 1967). See also Margo Vicedo, *The Nature and Nurture of Love: From Imprinting to Attachment in Cold War America* (University of Chicago Press, 2013), 1.
31 Hardyment, *Perfect Parents*, 223–7.
32 Stanley Coopersmith, *The Antecedents of Self-Esteem* (Consulting Psychologists Press, 1981), 4.
33 Julia Grant, *Raising Baby by the Book: The Education of American Mothers* (Yale University Press, 1998), 201. See also Rima D. Apple, *Mothers and Medicine: Social History of Infant Feeding, 1890–1950* (University of Wisconsin Press, 1988); Rima D. Apple and Janet Golden (eds.), *Mothers and Motherhood: Readings in American History* (Ohio State University Press, 1997).
34 Shirley C. Samuels, *Enhancing Self-Concept in Early Childhood* (Human Sciences Press, 1977), 16.
35 Anna Freud, *Normality and Pathology in Childhood: Assessments of Development* (International Universities Press, 1965).
36 Morris Rosenberg, "Parental Interest and Children's Self-Conceptions," *Sociometry*, 26 (1963), 35.
37 Benjamin Spock, *Dr. Spock's Baby and Child Care*, 8th ed. (Simon and Schuster, 2004 [1946]), 388–9.
38 Dorothy Corkville Briggs, *Your Child's Self-Esteem: The Key to His Life* (Doubleday & Co., 1970), 9.
39 Samuels, *Enhancing Self-concept*, 73.
40 Clarke, *Self-Esteem*, 32–3.
41 John E. Mack and Steven L. Ablon, *The Development and Sustenance of Self-Esteem in Childhood* (International Universities Press, 1983), 13.
42 Clarke, *Self-Esteem*, 32–3.
43 Jennifer Terry, "'Momism' and the Making of Treasonous Homosexuals," in Molly Ladd-Taylor and Lauri Umansky (eds.), *"Bad" Mothers: The Politics of Blame in Twentieth-Century America* (New York University Press, 1998), 174–5.
44 Maureen Stout, *The Feel-Good Curriculum: The Dumbing Down of America's Kids in the Name of Self-Esteem* (Perseus Books, 2000), 16.
45 Harry Hendrick, *Narcissistic Parenting in an Insecure World: A History of*

Parenting Culture, 1920s to the Present (Polity, 2016), 6; Cederström and Spicer, *Wellness Syndrome*; Ehrenreich, *Smile or Die*; Kathleen W. Jones, *Taming the Troublesome Child: American Families, Child Guidance, and the Limits of Psychiatric Authority* (Harvard University Press, 1999), 175.

46 Briggs, *Your Child's Self-Esteem*, xiii.
47 Briggs, *Your Child's Self-Esteem*, 1.
48 Briggs, *Your Child's Self-Esteem*, 2.
49 Briggs, *Your Child's Self-Esteem*, 3.
50 Briggs, *Your Child's Self-Esteem*, 3.
51 Briggs, *Your Child's Self-Esteem*, 3–4.
52 Mack and Ablon, *Development and Sustenance of Self-Esteem*, 12.
53 Mack and Ablon, *Development and Sustenance of Self-Esteem*, 17.
54 Hardyment, *Perfect Parents*, 299; Teri Chettiar, "Treating Marriage as 'The Sick Entity': Gender, Emotional Life and the Psychology of Marriage Improvement in Post-war Britain," *History of Psychology*, 18 (2015); David J. Bredehoft and Richard N. Hey, "An Evaluation Study of Self-Esteem: A Family Affair," *Family Relations*, 34 (1985).
55 E.N. Rexford, L.W. Sander, and T. Shapiro, "Introduction," in E.N. Rexford, L.W. Sander, and T. Shapiro (eds.), *Infant Psychiatry: A New Synthesis* (Yale University Press, 1976), xvii; see also Donald W. Winnicot, *The Maturational Process and the Facilitating Environment* (International Universities Press, 1976).
56 Frances Givelber, "The Parent–Child Relationship and the Development of Self-Esteem," in Mack and Ablon (eds.), *Development and Sustenance of Self-Esteem*.
57 Clarke, *Self-Esteem*, iv.
58 James T. Patterson, *Freedom Is Not Enough: The Moynihan Report and America's Struggle over Black Family Life from LBJ to Obama* (Basic Books, 2010), xii, 6.
59 Patterson, *Freedom Is Not Enough*, 34–5.
60 Daniel Patrick Moynihan, *The Moynihan Report: The Case for National Action* (Department of Labour, 1965), 5.
61 Moynihan, *Moynihan Report*, 6–12.
62 Moynihan, *Moynihan Report*, 29.
63 Moynihan, *Moynihan Report*, 47.
64 L. Alex Swan, "A Methodological Critique of the *Moynihan Report*," *Black Scholar*, 5 (1974), 18–21.
65 Patterson, *Freedom Is Not Enough*, 61.
66 Patterson, *Freedom Is Not Enough*, 76–8.
67 Taylor, *From #Blacklivesmatter to Black Liberation*, 40.

68 See, for example, Alan S. Berger and William Simon, "Black Families and the *Moynihan Report*: A Research Evaluation," *Social Problems*, 22 (1974).
69 Robert Hill, *The Strengths of Black Families* (Emerson Hall, 1972).
70 Leonard Lieberman, "The Emerging Model of the Black Family," *International Journal of Sociology of the Family*, 3 (1973).
71 Andrew Billingsley, *Black Families in White America* (Prentice-Hall, 1968), 75–8.
72 Rosenberg and Simmons, *Black and White Self-Esteem*, 74, 83.
73 McAdoo and McAdoo, *Black Children*, 10.
74 Joyce A. Ladner, *Tomorrow's Tomorrow: The Black Woman* (Doubleday, 1971), 49.
75 Lee Rainwater, *Behind Ghetto Walls: Black Families in a Federal Slum* (Allen Lane, 1970), 218.
76 Barry Silverstein and Ronald Krate, *Children of the Dark Ghetto: A Developmental Psychology* (Praeger, 1975), 14, 19. See also Andrew Billingsley and Jeanne M. Giovannoni, *Children of the Storm: Black Children and American Child Welfare* (Harcourt Brace Jovanovich, 1972).
77 McAdoo and McAdoo, *Black Children*, 9.
78 Gloria J. Powell and Marielle Fuller, *Black Monday's Children: A Study of the Effects of School Desegregation on Self-Concepts of Southern Children* (Appleton-Century-Crofts, 1973), 39.
79 Powell and Fuller, *Black Monday's Children*, 40, 306.
80 Poussaint, *Why Blacks Kill Blacks*, 103.
81 Poussaint, *Why Blacks Kill Blacks*, 103–4.
82 Poussaint, *Why Blacks Kill Blacks*, 105.
83 Algae O. Harrison, "The Black Family's Socializing Environment: Self-Esteem and Ethnic Attitude among Black Children," in McAdoo and McAdoo, *Black Children*, 186, 189.
84 James P. Comer and Alvin F. Poussaint, *Raising Black Children* (Plume, 1992 [1975]), 1.
85 Comer and Poussaint, *Raising Black Children*, 2, 4, 16.
86 Comer and Poussaint, *Raising Black Children*, 60–1.
87 Phyllis Harrison-Ross and Barbara Wyden, *The Black Child: A Parent's Guide* (Peter H. Wyden, 1973), xxvii.
88 Harrison-Ross and Wyden, *The Black Child*, xxiii–xxiv.
89 Amos N. Wilson, *The Developmental Psychology of the Black Child* (Africana Research Publications, 1978), 6–7.
90 Marie Ferguson Peters, "Racial Socialization of Young Black Children," in McAdoo and McAdoo, *Black Children*, 164.

91 Mikki Kendall, *Hood Feminism: Notes from the Women that a Movement Forgot* (Penguin, 2020), 240–1.
92 Peters, "Racial Socialization of Young Black Children," 171.
93 Robert C. Johnson, "The Black Family and Black Community Development," *Journal of Black Psychology*, 8 (1981), 30, 36.
94 Franklin and Franklin, "A Psychoeducational Perspective," 209.

Chapter 5: School

1 Michele and Craig Borba, *Self-Esteem: A Classroom Affair. 101 Ways to Help Children Like Themselves* (Winston Press, 1978), 3.
2 Winters, *Black Fatigue*, 12–13.
3 Michael Eric Dyson, *Tears We Cannot Stop: A Sermon to White America* (St. Martin's Press, 2017), 126–9.
4 Ronald Geraty, "Education and Self-Esteem," in Mack and Ablon (eds.), *Development and Sustenance of Self-Esteem*, 263.
5 Stephen Vassalo, *Neoliberal Selfhood* (Cambridge University Press, 2021), 1–3.
6 Paula S. Fass and Michael Grossberg (eds.), *Reinventing Childhood after World War II* (University of Pennsylvania Press, 2012), xiii.
7 McGirr, *Suburban Warriors*, 179–81.
8 John L. Rury, *Education and Social Change: Themes in the History of American Schooling* (Lawrence Erlbaum, 2005), 150–1, 187.
9 Borba and Borba, *Self-Esteem*, 2.
10 Pauline S. Sears and Vivian S. Sherman, *In Pursuit of Self-Esteem: Case Studies of Eight Elementary School Children* (Wadsworth, 1964), 2.
11 For an overview, see Diane Ravitch, *The Troubled Crusade: American Education, 1945–1980* (Basic Books, 1983).
12 Stout, *The Feel-Good Curriculum*, 76.
13 Bergeron, *Challenging the Cult of Self-Esteem*, xviii.
14 Jack Martin, "A Case against Heightened Self-Esteem as an Educational Aim," *Journal of Thought*, 42 (2007), 55.
15 Cruikshank, *Will to Empower*, 234.
16 Bergeron, *Challenging the Cult of Self-Esteem*, 99.
17 Bergeron, *Challenging the Cult of Self-Esteem*, 36–7
18 Jack Martin and Ann-Marie McLellan, *The Education of Selves: How Psychology Transformed Students* (Oxford University Press, 2013).
19 Cecil H. Patterson, *Humanistic Education* (Prentice-Hall, 1973), x, 13, 20–1.
20 Nathaniel Lees Gage, *Educational Psychology* (Rand McNally College, 1975).
21 Robert B. Burns, *Self-Concept Development and Education* (Holt, Rinehart, and Winston, 1982), v–vi.

22 Prescott Lecky, *Self-Consistency: A Theory of Personality* (Island Press, 1945).
23 E.g. Martin B. Fink, "Self-Concept as it Relates to Academic Underachievement," *California Journal of Educational Research*, 13 (1962); Robert L. Williams and Spurgeon Cole, "Self-Concept and School Adjustment," *Personnel and Guidance Journal*, 46 (1968).
24 George W. Wallis, "Self-Esteem and Classroom Standards," *Pedagogy Journal of Education*, 47 (1969), 143.
25 William P. Ferris, "Raising Student Self-Esteem (For a Change)," *English Journal*, 60 (1971), 379.
26 Ferris, "Raising Student Self-Esteem," 379.
27 E.g. Geraty, "Education and Self-Esteem," 255–6.
28 Nancy F. Sizer and Theodore R. Sizer (eds.), *Moral Education: Five Lectures* (Harvard University Press, 1970).
29 James A. Beane and Richard P. Lipka, *Self-Concept, Self-Esteem, and the Curriculum* (Allyn and Bacon, 1984), 18.
30 Burns, *Self-Concept Development*, 201.
31 William Watson Purkey, *Self-Concept and School Achievement* (Prentice Hall, 1970), 20, 43, 50, 51–2, 55.
32 Robert Leonetti, *Self-Concept and the School Child: How to Enhance Self-Confidence and Self-Esteem* (Philosophical Library, 1980), 18.
33 Leonetti, *Self-Concept and the School Child*, 19–21, 39–44, 51–8.
34 Edward Zigler, "The Effectiveness of Head Start: Another Look," *Educational Psychologist*, 13 (1978); "America's Head Start Program: An Agenda for Its Second Decade," *Young Children*, 33 (1978).
35 Sanford Golin, Eugene Davis, Edward Zuckerman, and Nelson Harrison, "Psychology in the Community: Project Self-Esteem," *Psychological Reports*, 26 (1970); Sanford Golin, *Project Self-Esteem: Some Effects of an Elementary School Black Studies Program* (National Institute of Mental Health, 1971).
36 Sandy McDaniel and Peggy Bielen, *Project Self-Esteem: A Parent Involvement Program for Elementary-Age Children* (B.L. Winch, 1986), x–xi.
37 Frederick D. Harper, "Developing a Curriculum of Self-Esteem for Black Youth," *Journal of Negro Education*, 46 (1977).
38 Donald E. Michel and Dorothea Martin Farrell, "Music and Self-Esteem: Disadvantaged Problem Boys in an All-Black Elementary School," *Journal of Research in Music Education*, 21 (1973).
39 Joanne Hendrick, *The Whole Child: New Trends in Early Education* (C.V. Mosby, 1975), 103–4.
40 Borba and Borba, *Self-Esteem*, Chs. 2 and 3.
41 Beane and Lipka, *Self-Concept*, 91–9.

42 James A. Beane and Richard P. Lipka, *When the Kids Come First: Enhancing Self-Esteem* (National Middle Schools Association, 1987), 31.
43 Jack Canfield and Harold C. Wells, *100 Ways to Enhance Self-Concept in the Classroom: A Handbook for Teachers and Parents* (Prentice Hall, 1976).
44 Purkey, *Inviting School Success*, 10.
45 Burns, *Self-Concept Development*, 251.
46 Verne Faust, *Self Esteem in the Classroom* (Thomas Paine, 1980), 55.
47 Beane and Lipka, *When the Kids Come First*, 5.
48 Purkey, *Inviting School Success*, 28–38.
49 Geraty, "Education and Self-Esteem," 265–6.
50 "Bolstering Self-Esteem to Reduce Pupils' Stress," *New York Times* (May 17, 1992), WC16.
51 "Workshops that Lift Self-Esteem," *New York Times* (January 29, 1989), 10.
52 Burns, *Self-Concept Development*, 255.
53 Stout, *Feel-Good Curriculum*, 3.
54 Stout, *Feel-Good Curriculum*, 142.
55 Donald H. Smith, "The Black Revolution and Education," in James A. Banks and Jean D. Grambs (eds.), *Black Self-Concept: Implications for Education and Social Science* (McGraw Hill, 1972), 44–5.
56 Smith, "The Black Revolution," 52.
57 William C. Kvaraceus (ed.), *Negro Self-Concept: Implications for School and Citizenship* (Tufts University, 1964).
58 Banks and Grambs (eds.), *Black Self-Concept*, xii.
59 James A. Banks, "Racial Prejudice and the Black Self-Concept," in Banks and Grambs (eds.), *Black Self-Concept*, 5–35.
60 Banks, "Racial Prejudice," 26.
61 Roy A. Weaver, *School Organization and Black Children's Esteem and Future-Focused Role Images* (R. & E. Research Associates, 1979).
62 Toni Morrison, *The Bluest Eye* (Vintage, 2019 [1970]), x–xi.
63 Morrison, *The Bluest Eye*, 18.
64 Morrison, *The Bluest Eye*, 21.
65 Stuart Buck, *Acting White: The Ironic Legacy of Desegregation* (Yale University Press, 2010), 3–4.
66 Robert Coles, *Children of Crisis: A Study of Courage and Fear* (Atlantic Monthly, 1967), 153.
67 Winters, *Black Fatigue*, 12.
68 Theoharis and Theoharis, *These Yet to Be United States*, 193.
69 In *White Fragility*, Robin Diangelo writes at length about self-confessed white nonracists still perpetuating a system that works against Blacks.

70 Hacker, *Two Nations*, 171; Baughman, *Black Americans*, 31–2; Gordon D. Morgan, "A Brief Overview of Ghetto Education: Elementary to College," in Edgar G. Epps, *Black Students in White Schools* (Charles A. Jones Publishing Co., 1972), 8; Vivian Gunn Morris and Curtis L. Morris, *The Price they Paid: Desegregation in an African American Community* (Teachers College Press, 2002), 78–9.

71 Derrick Bell, *And We Are Not Saved: The Elusive Quest for Racial Justice* (Basic Books, 1987), 108–9.

72 Derrick Bell, *Silent Covenants: Brown v. Board of Education and the Unfulfilled Hopes for Racial Reform* (Oxford University Press, 2004), 96–7, 3–4.

73 Bell, *Silent Covenants*, 112–13.

74 Carol Anderson, *White Rage: The Unspoken Truth of our Racial Divide* (Bloomsbury, 2020 [2016]), 120–1.

75 Austin Channing Brown, *I'm Still Here: Black Dignity in a World Made for Whiteness* (Convergent, 2018), 44.

76 Brown, *I'm Still Here*, 44–5.

77 Bell, *And We Are Not Saved*, 110.

78 Harper, "Developing a Curriculum," 134.

79 Harper, "Developing a Curriculum," 135, 140.

80 Comer and Poussaint, *Raising Black Children*, 178, 212.

81 Comer and Poussaint, *Raising Black Children*, 120.

82 Jawanza Kunjufu, *Developing Positive Self-Images in Black Children* (Africa-American Images, 1984), vii, ix, x.

83 Kunjufu, *Developing Positive Self-Images*, 3.

84 Kunjufu, *Developing Positive Self-Images*, 6. Kunjufu quotes here from Pansye Atkinson and Fred Hord, *Save the Children* (Frostburg: Office of Minority Affairs, 1983).

85 Kunjufu, *Developing Positive Self-Images*, 8, 9, 29, 98.

86 Janice E. Hale-Benson, *Black Children: Their Roots, Culture and Learning Styles*, 2nd ed. (Johns Hopkins University Press, 1982), xi–xii, 1.

87 Hale-Benson, *Black Children*, 3.

Chapter 6: Teenagers

1 Jon Savage, *Teenage: The Creation of Youth Culture* (Chatto & Windus, 2007); James Gilbert, *A Cycle of Outrage: America's Reaction to the Juvenile Delinquent in the 1950s* (Oxford University Press, 1986); McGirr, *Suburban Warriors*; Heather Munro Prescott, *A Doctor of their Own: The History of Adolescent Medicine* (Harvard University Press, 1998).

2 Cavallo, *Fiction of the Past*, Ch. 3.
3 Jean Twenge, *Generation Me* (Free Press, 2006), 136.
4 Michal Shapira, *The War Inside: Psychoanalysis, Total War, and the Making of the Democratic Self in Postwar Britain* (Cambridge University Press, 2013), 17–18, 200, 84, 238.
5 Granville Stanley Hall, *Adolescence: Its Psychology and Its Relations to Physiology, Anthropology, Sociology, Sex, Crime, Religion and Education* (Appleton, 1904).
6 Anna Freud, *Adolescence as a Developmental Disturbance* (Basic Books, 1969). For discussion, see Alex Holder, *Anna Freud, Melanie Klein and the Psychoanalysis of Children and Adolescents* (Routledge, 2018 [2005]).
7 Erik Erikson, *Identity: Youth and Crisis* (W.W. Norton & Co., 1968).
8 Carl Wells and Irving R. Stuart (eds.), *Self-Destructive Behaviour in Children and Adolescents* (Van Nostrand Reinhold, 1981), xi.
9 Janet Kizziar and Judy Hagedorn, *Search for Acceptance: The Adolescent and Self-Esteem* (Nelson-Hall, 1979), 7.
10 Ian Miller, "Ending the 'Cult of the Broken Home': Divorce, Children and the Changing Emotional Dynamics of Separating British Families, c.1945–90," *Twentieth-Century British History*, 32 (2021).
11 For discussion, see Michael Grossberg, "Liberation and Caretaking: Fighting over Children's Rights in Postwar America," in Fass and Grossberg (eds.), *Reinventing Childhood*.
12 Jerome B. Dusek, "The Development of Self-Concept during the Adolescent Years," *Monographs of the Society for Research in Child Development*, 46 (1981).
13 Mack and Ablon, *Development and Sustenance of Self-Esteem*, 17.
14 Morris Rosenberg, *Society and the Adolescent Self-Image* (Princeton University Press, 1965), 3–4.
15 "How to Succeed as a Teenager," *New York Times* (April 18, 1965), SM94.
16 Robert G. Simmons, Florence Rosenberg, and Morris Rosenberg, "Disturbance in the Self-Image at Adolescence," *American Sociological Review*, 38 (1973).
17 Ian Miller, "Bedwetting, Doctors and the Problematization of Youth in Britain and America, c.1800–1980," *Journal of Social History*, 52 (2019).
18 Miller, "Ending the 'Cult of the Broken Home'."
19 Daniel Offer, Richard C. Marohn, and Eric Ostrov, *The Psychological World of the Juvenile Delinquent* (Basic Books, 1979), 7.
20 E.g. Robert H. Abramowitz, Anne C. Peterson, and John E. Schulenberg, "Changes in Self-Image during Adolescence," in Daniel Offer, Eric Ostrov, and Kenneth I. Howard (eds.), *Patterns of Adolescent Self-Image* (Jossey-Bass, 1984).
21 Howard B. Kaplan, *Deviant Behavior in Defense of Self* (Academic Press, 1980); *Self-Attitudes and Deviant Behavior* (Goodyear, 1975).

22 E.g. John D. McCarthy and Dean R. Hope, "The Dynamics of Self-Esteem and Delinquency," *American Journal of Sociology*, 90 (1984); W. Alex Mason, "Self-Esteem and Delinquency Revisited (Again): A Test of Kaplan's Self-Derogation Theory of Delinquency Using Latent Growth Curve Modeling," *Journal of Youth and Adolescence*, 30 (2001).

23 For discussion, see Jerome B. Fischer and Carl A. Bersarni, "Self-Esteem and Institutionalized Delinquent Offenders: The Role of Background Characteristics," *Adolescence*, 14 (1979).

24 Paula S. Fass, "The Child-Centered Family? New Rules in Postwar America," in Fass and Grossberg (eds.), *Reinventing Childhood*, 6–7.

25 Kizziar and Hagedorn, *Search for Acceptance*, 1; James Battle, *9 to 19: Crucial Years for Self-Esteem in Children and Youth* (James Battle, 1990), 39, 59; Kenneth L. Jones, Louis W. Shainberg, and Curtis O. Byer, *Drugs, Alcohol and Tobacco* (Canfield Press, 1970), 41–2.

26 Spencer A. Rathus and Larry J Siegel, "Delinquent Attitudes and Self-Esteem," *Adolescence*, 8 (1973).

27 Howard Kaplan, "Self-Esteem and Self-Derogation Theory of Drug Abuse," in Dan J. Lettieri, Mollie Sayers, and Helen Wallenstein Pearson (eds.), *Theories on Drug Abuse: Selected Contemporary Perspectives* (National Institute on Drug Abuse, 1980); see also Kizziar and Hadedorn, *Search for Acceptance*, 3.

28 Kyle Riismandel, "'Say You Love Satan': Teens and Popular Occulture in 1980s America," in Susan Eckelmann Berghel, Sara Fieldston, and Paul M. Renfro (eds.), *Growing Up in America: Youth and Politics since 1945* (University of Georgia Press, 2019).

29 Cynthia Tennant-Clark, Janet J. Fritz, and Fred Beauvais, "Occult Participation: Its Impact on Adolescent Development," *Adolescence*, 24 (1989), 759.

30 Marcia R. Rudin, "Cults and Satanism: Threats to Teens," *NASSP Bulletin* (1 May 1990).

31 Tennant-Clark, Fritz, and Beauvais, "Occult Participation."

32 Rudin, "Cults and Satanism."

33 Michael D. Langone and Linda O. Blood, *Satanism and Occult-Related Violence: What You Should Know* (International Cultic Studies Association, 1990), 39.

34 Gayland W. Hurst and Robert L. Marsh, *Satanic Cult Awareness* (AR, 1993), 30.

35 Langone and Blood, *Satanism and Occult-Related Violence*, 12.

36 Vicki L. Dawkins, *Devil Child: A Terrifying True Story of Satanism and Murder* (St. Martin's Press, 1989).

37 Saul V. Levine, "Life in the Cults," in Marc Galanter (ed.), *Cults and New Religious Movements: A Report of the American Psychiatric Association* (American Psychiatric Association, 1989), 101.
38 James T. Richardson, Joel Best, and David G. Bromley (eds.), *The Satanism Scare* (Routledge, 1991), 21, 178, 276; Cynthia M. Clark, "Clinical Assessment of Adolescents Involved in Satanism," *Adolescence*, 29 (1994).
39 Jeffrey S. Victor, "The Extent of Satanic Crime is Exaggerated," in Tamara L. Roleff (ed.), *Satanism* (Greenhaven, 2002), 42.
40 Langone and Blood, *Satanism and Occult-Related Violence*, 56–7.
41 Twenge, *Generation Me*.
42 Alex Mold and Virginia Berridge, *Voluntary Action and Illegal Drugs: Health and Society in Britain since the 1960s* (Palgrave Macmillan, 2010), 23–4, 38.
43 Robert B. Millman, Elizabeth T. Khuri, and David Hammond, "Perspectives on Drug Use and Abuse," in Wells and Stuart (eds.), *Self-Destructive Behavior*, 131.
44 For discussion, see Ronald A. Steffenhagen and Jeff D. Burns, *The Social Dynamics of Self-Esteem: Theory to Therapy* (Praeger, 1987), 53.
45 Frank Maguire and Stephen Wilson, "Self-Esteem and Subjective Effects during Marijuana Intoxication," *Journal of Drug Issues*, 15 (1985).
46 Thomas J. Glynn, Carl G. Leukefeld, and Jacqueline P. Ludford (eds.), *Preventing Adolescent Drug Abuse: Intervention Strategies* (Department of Health and Human Services, 1983), 119.
47 Riismandel, "'Say You Love Satan'," 213–14.
48 "Suburbs Try Their Own Kind of Drug War in the Schools," *New York Times* (September 24, 1989), E7.
49 Paula Englander-Golden, Joan Elconin Jackson, Karen Crane, et al., "Communication Skills and Self-Esteem in Prevention of Destructive Behaviors," *Adolescence*, 24 (1989), 481.
50 See, e.g. Carl May, "Perceptions of Self, Self-Esteem and the Adolescent Smoker," *Health Education Journal*, 58 (1999).
51 Gillian N. Penny and James O. Robinson, "Psychological Resources and Cigarette Smoking in Adolescents," *British Journal of Psychology*, 77 (1986).
52 Frank F. Furstenberg, *Destinies of the Disadvantaged: The Politics of Teen Childbearing* (Russell Sage Foundation, 2007).
53 Lucie Rudd, "Pregnancies and Abortions," in Wells and Stuart (eds.), *Self-Destructive Behaviour*, 212.
54 Barbara Halpren Elkes and John A. Crocitto, "Self-Concept of Pregnant Adolescents: A Case Study," *Journal of Humanistic Education and Development*, 25 (1987).

55 Virginia Abernethy, "Illegitimate Conception among Teenagers," *American Journal of Public Health*, 64 (1974).
56 Cynthia Robbins, Howard B. Kaplan, and Steven S. Martin, "Antecedents of Pregnancy among Unmarried Adolescents," *Journal of Marriage and the Family*, 47 (1985).
57 C.E. Zongker, "Self-Concept Differences between Single and Married School Age Mothers," *Journal of Youth and Adolescence*, 9 (1980).
58 For discussion, see Rachel B. Robinson and Deborah I. Frank, "The Relation between Self-Esteem, Sexual Activity and Pregnancy," *Adolescence*, 29 (1994).
59 Roy F. Baumeister, *Escaping the Self: Alcoholism, Spirituality, Masochism and Other Flights from the Burden of Selfhood* (Basic Books, 1991), 12.

Chapter 7: Girls and Women
1 Michael Gurian, *The Wonder of Girls: Understanding the Hidden Nature of Our Daughters* (Atria, 2002), 191.
2 Herman, *Romance of American Psychology*, 312; Katie Wright, *Rise of the Therapeutic Society*; Aubry and Travis (eds.), *Rethinking Therapeutic Culture*.
3 Hannah Dudley-Shotwell, *Revolutionizing Women's Healthcare: The Feminist Self-Help Movement in America* (Rutgers University Press, 2020), 102–3.
4 Peggy White and Starr Goode, "The Small Group in Women's Liberation," *Women: A Journal of Liberation*, 1 (1969), 56–7.
5 Nathaniel Branden, *A Woman's Self-Esteem: Stories of Struggle, Stories of Triumph* (Nash, 2001 [1969]).
6 Ruth C. Wylie, *The Self-Concept*, vol. II, 2nd ed. (University of Nebraska Press, 1979 [1961]), 243.
7 See, e.g., Phyllis Chesler, *Women and Madness* (Avon Books, 1973); Angela Phillips, *Our Bodies Ourselves: A Health Book by and for Women* (Allen Lane, 1978); Barbara Ehrenreich and Deirdre English, *For Her Own Good: Two Centuries of the Experts' Advice to Women* (Pluto Press, 1979); Elaine Showalter, *The Female Malady: Women, Madness and English Culture, 1830–1980* (Virago, 1987).
8 See, among others, Juliet Mitchell, *Psychoanalysis and Feminism* (Penguin, 1990).
9 See Moskowitz, *In Therapy We Trust*, 209; Herman, *Romance of American Psychology*, 291.
10 Lucy Delap, *Feminisms: A Global History* (University of Chicago Press, 2020); Alice Echols, *Daring to be Bad: Radical Feminism in America, 1967–1975* (University of Minnesota Press, 1989), 243–5.
11 Pauline Rose Clance and Suzanne Ament Imes, "The Imposter Phenomenon

in High-Achieving Women: Dynamics and Therapeutic Intervention," *Psychotherapy: Theory, Research and Practice*, 15 (1978).
12 Carol Eagle and Carol Colman, *Your Adolescent Daughter: How to Encourage your Daughter to Maintain her Self-Esteem and Reach her Full Potential* (Piatkus, 1995), 12.
13 Mary Pipher, *Reviving Ophelia: Saving the Selves of Adolescent Girls* (Grosset/Putnam, 1994), 12–13, 27, 149–50.
14 Richard H. Klemer, "Self-Esteem and College Dating Experience as Factors in Mate Selection and Marital Happiness: A Longitudinal Study," *Journal of Marriage and Family*, 33 (1971).
15 For a nuanced discussion of these debates, see Ali Haggett, *Desperate Housewives, Neuroses and the Domestic Environment, 1945–1970* (Pickering and Chatto, 2012).
16 Anne Statham Macke, George W. Bohrnstedt, and Ilene N. Bernstein, "Housewives' Self-Esteem and their Husbands' Success: The Myth of Vicarious Involvement," *Journal of Marriage and Family*, 41 (1979).
17 Linda Tschirhart Sanford and Mary Ellen Donovan, *Women and Self-Esteem* (Penguin, 1984), xiv–xv.
18 Sanford and Donovan, *Women and Self-Esteem*, 3–4, 6, 12.
19 Steinem, *Revolution from Within*, 9.
20 Steinem, *Revolution from Within*, 29–30, 9–10, 30.
21 Steinem, *Revolution from Within*, 25.
22 Steinem, *Revolution from Within*, 83.
23 Carolynn Hillman, *Recovery of Your Self-Esteem: A Guide for Women* (Simon and Schuster, 1992), 12.
24 Hillman, *Recovery of Your Self-Esteem*, 17–18, 39.
25 Elayne Rapping, *The Culture of Recovery: Making Sense of the Self-Help Movement in Women's Lives* (Beacon Press, 1996), 6, 161.
26 See, among others, Shapira, *War Inside*.
27 Deirdre English, "Review: Gloria Steinem, *Revolution from Within*," *New York Times* (February 2, 1992).
28 Cloud, *Control and Consolation*, 105–14, 129–30.
29 Susan Faludi, *Backlash: The Undeclared War against Women* (Chatto and Windus, 1991), 371–3.
30 Susan J. Douglas, *Where the Girls Are* (Penguin, 1994), 246.
31 Jessa Crispin, *Why I Am Not a Feminist: A Feminist Manifesto* (Melville House, 2017), 9.
32 Crispin, *Why I Am Not a Feminist*, 147.
33 Kathy Peiss, *Hope in a Jar: The Making of America's Beauty Culture* (Henry Holt, 1998), 260–1.

34 Gerald R. Adams and Sharyn M. Crossman, *Physical Attractiveness: A Cultural Imperative* (Libra, 1978).
35 Cited in J.C. Barden "The Hazards of Being Beautiful: Basis of Self-Esteem Men's Viewpoint," *New York Times* (May 2, 1974).
36 "The Un-Beautiful People," *New York Times* (March 18, 1976).
37 "Not Just a Pretty Face," *New York Times* (August 26, 1984).
38 "Woman's Image in a Mirror," *New York Times* (February 6, 1992).
39 Nicolas Rasmussen, *Fat in the Fifties: America's First Obesity Crisis* (Johns Hopkins University Press, 2019).
40 Louise Foxcroft, *Calories and Corsets: A History of Dieting over 2,000 Years* (Profile Books, 2011), 195.
41 Rita Freedman, *Bodylove: Learning to Like Our Looks and Ourselves: A Practical Guide for Women* (Gürze, 2002 [1988]), 129; Roberta Pollock, *Never Too Thin: Why Women are at War with Their Bodies* (Prentice Hall, 1989), 26.
42 Naomi Wolf, *The Beauty Myth: How Images of Beauty Are Used Against Women* (Vintage Books, 2015 [1990]), 2–3.
43 Peter Stearns, *Fat History: Bodies and Beauty in the Modern West* (New York University Press, 1997), 98; Hesse-Biber, *Am I Thin Enough Yet?*, 28, 102–6.
44 Susie Orbach, *Fat is a Feminist Issue I & II: The Anti-Diet Guide for Women* (Arrow, 1998 [1978]), 24–5, 124, 133.
45 E.g. Jean Jacobs Brumberg, *Fasting Girls: The Emergence of Anorexia Nervosa as a Modern Disease* (Harvard University Press, 1988), 254.
46 Susie Orbach, *Hunger Strike: The Anorectic's Struggle as a Metaphor for our Age* (Avon, 1986), 13–14, 73–4, 145–6, 204.
47 Mary Pipher, *Hunger Pains: The Modern Woman's Tragic Quest for Thinness* (Ballantine, 1995), 2–5, 117.
48 Mary Ann Mayo, *Looking Good but Feeling Bad* (Servant Publications, 1992), 11–13, 14, 15.
49 Freedman, *Bodylove*, 21–3, 41, 61, 286.
50 Julie A. Hollyman, J. Hubert Lacey, P.J. Whitfield, and J.S.P. Wilson, "Surgery for the Psyche: A Longitudinal Study of Women Undergoing Reduction Mammoplasty," *British Journal of Plastic Surgery*, 39 (1986); Sander Gilman, *Making the Body Beautiful: A Cultural History of Aesthetic Surgery* (Princeton University Press, 2001).
51 Jane Wegscheider and Esther R. Rome, *Sacrificing Ourselves for Love: Why Women Compromise Health and Self-Esteem … and How to Stop* (Crossing Press, 1996), 7, 31–2, 73–6, 123.
52 Karen Blaker, *Born to Please* (St. Martin's Press, 1990), 2.
53 Wendy Chapkis, *Beauty Secrets* (Women's Press, 1986); Faludi, *Backlash*, 240.

54 Elizabeth Haiken, *Venus Envy: A History of Cosmetic Surgery* (Johns Hopkins University Press, 1997), 9–10; Peiss, *Hope in a Jar*, 4; Virginia L. Blum, *Flesh Wounds: The Culture of Cosmetic Surgery* (University of California Press, 2003), 51; Kathy Davis, *Reshaping the Female Body: The Dilemma of Cosmetic Surgery* (Routledge, 1995), 4–5, 49.
55 Blum, *Flesh Wounds*, 63, 294; see also Peiss, *Hope in a Jar*, 268.
56 Elizabeth Taylor, *Elizabeth Takes Off* (Macmillan, 1988), 2, 21.
57 Carl Elliott, *Better than Well: American Medicine Meets the American Dream* (W.W. Norton & Co., 2003), 203.
58 I borrow here from Haiken, *Venus Envy*, 283.
59 *Shortchanging Girls, Shortchanging America* (American Association of University Women, 1991), 6, 8.
60 *Shortchanging Girls*, 4–5.
61 Cited in Suzanne Daley, "Little Girls Lose their Self-Esteem on Way to Adolescence," *New York Times* (January 9, 1991).
62 Myra Sadker and David Sadker, *Failing at Fairness: How America's Schools Cheat Girls* (Charles Scribner's Sons, 1994), 1.
63 Carol Gilligan, *In a Different Voice: Psychological Theory and Women's Development* (Harvard University Press, 1982).
64 Cited in Daley, "Little Girls Lose their Self-Esteem."
65 Lyn Mikel Brown and Carol Gilligan, *Meeting at the Crossroads: Women's Psychology and Girls' Development* (Harvard University Press, 1992), 4–5.
66 "Girls and Puberty," *New York Times* (November 4, 1997).
67 Emily Hancock, *The Girl Within: A Radical New Approach to Female Identity* (Pandora, 1990), 245, 258; see also Jane E. Brody, "Personal Health; Girls and Puberty: The Crisis Years," *New York Times* (November 4, 1997).
68 Peggy Orenstein, *Schoolgirls: Young Women, Self-Esteem and the Confidence Gap* (Anchor Books, 1995), xix, xxii, xxxi.
69 These criticisms were later voiced in Rosalind Barnett and Caryl Rivers, *Same Difference: How Gender Myths are Hurting our Relationships, Our Children and Our Jobs* (Basic Books, 2004), 236–7.
70 For discussion, see Marlene Mackie, "The Domestication of Self: Gender Comparisons of Self-Imagery and Self-Esteem," *Social Psychology Quarterly*, 46 (1983); Arlene Eskilson and Mary Glenn Wiley, "Parents, Peers, Perceived Pressure, and Adolescent Self-Concept: Is a Daughter a Daughter All of Her Life?," *Sociological Quarterly*, 28 (1987).
71 Wylie, *The Self-Concept*, 328.
72 Janet S. Hyde, "The Gender Similarities Hypothesis," *American Psychologist*, 60 (2005).

73 James Kenway and Sue Willis, "Self-Esteem and the Schooling of Girls: An Introduction," in James Kenway and Sue Willis (eds.), *Hearts and Minds: Self-Esteem and the Schooling of Girls* (London: Falmer Press, 1990), 2, 10–11.

74 Cited in "The Myth of Schoolgirls' Low Self-Esteem," *Wall Street Journal* (October 3, 1994).

75 Christina Hoff Sommers, *Who Stole Feminism? How Women Have Betrayed Women* (Simon and Schuster, 1994), 15, 25, 141, 151.

76 Susan Pinker, *The Sexual Paradox: Troubled Boys, Gifted Girls, and the Real Difference Between the Sexes* (Atlantic Books, 2008), 226–30.

77 Rosalind Barnett and Caryl Rivers, *Same Difference: How Gender Myths are Hurting our Relationships, Our Children and Our Jobs* (Basic Books, 2004), 227, 238.

78 Hyde, "The Gender Similarities Hypothesis."

79 Steven Pinker, *The Blank Slate: The Modern Denial of Human Nature* (BCA, 2002).

80 Gurian, *Wonder of Girls*.

81 Recent texts not already cited include Simon Baron-Cohen, *The Essential Difference: Men, Women and the Extreme Male Brain* (Allen Lane, 2003); Cordelia Fine, *Delusions of Gender: The Real Science behind Sex Differences* (Icon, 2010); Cordelia Fine, *Testosterone Rex: Unmaking the Myth of our Gendered Minds* (W.W. Norton, 2017); Daphna Joel and Luba Vikhanski, *Gender Mosaic: Beyond the Myth of the Male and Female Brain* (Endeavour, 2019).

82 Sadker and Sadker, *Failing at Fairness*, 78, 80–1, 98.

83 Barnett and Rivers, *Same Difference*, 244–5.

84 Dana Becker, *The Myth of Empowerment: Women and Therapeutic Culture in America* (New York University Press, 2005), 2–3, 136.

85 Jane Kenway, Sue Willis, and Jenny Nevard, "The Subtle Politics of Self-Esteem Programs for Girls," in Kenway and Willis (eds.), *Hearts and Minds*, 35.

86 Kenway, Willis, and Nevard, "The Subtle Politics of Self-Esteem Programs," 40.

87 Gurian, *Wonder of Girls*, 7.

88 Gurian, *Wonder of Girls*, 16.

89 Boyd, *In the Company of My Sisters*, 57–8.

90 See, among others, Alexander, *The New Jim Crow*; DiAngelo, *White Fragility*.

91 Alexandra Rutherford, *Psychology at the Intersections of Gender, Feminism, History and Culture* (Cambridge University Press, 2021), 24–5, 31.

92 Catherine Rottenberg, *The Rise of Neoliberal Feminism* (Oxford University Press, 2018), ix.

93 Rafia Zakaria, *Against White Feminism* (Hamish Hamilton, 2021), ix, x, 10, 100–1.
94 Boyd, *In the Company of My Sisters*, 58–9, 147.
95 bell hooks, *Ain't I a Woman: Black Women and Feminism* (Pluto Press, 1982), 1–12, 97, 124, 136–8.
96 hooks, *Ain't I a Woman*, 142–3.
97 hooks, *Ain't I a Woman*, 149.
98 Mikki Kendall, *Hood Feminism: Notes from the Women that a Movement Forgot* (Penguin, 2020), 240–1.

Chapter 8: A Self-Esteem Task Force

1 Nolan Jr., *The Therapeutic State*, 9, 14, 1–3.
2 Elizabeth Puttick, "Human Potential Movement," in Christopher Hugh (ed.), *Encyclopedia of New Religions* (Lion Hudson, 2006).
3 Walter Truett Anderson, *The Upstart Spring: Esalen and the Human Potential Movement: The First Twenty Years* (Addison-Wesley, 1983); Jeffrey J. Kripal, *Esalen: America and the Religion of No Religion* (University of Chicago Press, 2007).
4 "Now, the California Task Force to Promote Self-Esteem," *New York Times* (October 11, 1986).
5 Nick Totton, *Psychotherapy and Politics* (SAGE, 2000), 3–4.
6 Karl R. Popper, *Conjectures and Refutations: The Growth of Scientific Knowledge* (Routledge & Kegan Paul, 1963), 381–3; Robert Nozick, *Anarchy, State, and Utopia* (Basic Books, 1974), 245; John Rawls, *A Theory of Justice* (Harvard University Press, 1971), 440.
7 Joseph Kahne, "The Politics of Self-Esteem," *American Educational Research Journal*, 33 (1996), 6.
8 Nolan Jr., *The Therapeutic State*, 165.
9 *Toward a State of Esteem: The Final Report of the California Task Force to Promote Self-Esteem and Personal and Social Responsibility* (California State Department of Education, 1990), 26.
10 Neil J. Smesler, "Self-Esteem and Social Problems: An Introduction," in Andrew Mecca, Neil J. Smelser, and John Vasconcellos (eds.), *The Social Importance of Self-Esteem* (University of California Press, 1989), 1.
11 "Now, the California Task Force to Promote Self-Esteem," *New York Times* (October 11, 1986).
12 "Ministry for Feeling Good," *London Times* (January 22, 1988).
13 "Will the Real Self-Esteem Stand Up?" *New York Times* (November 9, 1988).
14 *Toward a State of Esteem*, 19–20, 21.

15 Kahne, "The Politics of Self-Esteem," 10.
16 Quoted in "Task Force to Ensure Californians Feel Good," *London Times* (March 3, 1987).
17 See https://www.theguardian.com/lifeandstyle/2017/jun/03/quasi-religious-great-self-esteem-con.
18 "Using Self-Esteem to Fix Society's Ills," *New York Times* (March 28, 1990).
19 *Toward a State of Esteem*, ix.
20 "Using Self-Esteem to Fix Society's Ills."
21 "A Study on Feelings Creates Bad Ones," *New York Times* (February 20, 1990); *Toward a State of Esteem*, 156–61.
22 *Toward a State of Esteem*, vii–viii.
23 *Toward a State of Esteem*, ix
24 *Toward a State of Esteem*, x, 26.
25 *Toward a State of Esteem*, 22, 59, 28–9.
26 *Toward a State of Esteem*, 61, 22–3.
27 E.g. Kahne, "The Politics of Self-Esteem," 12.
28 Smesler, "Self-Esteem and Social Problems," 18, 21, 23.
29 See Bonnie Bahtti, David Derezotes, Seung-Ock Kim, and Harry Specht, "The Association between Child Maltreatment and Self-Esteem"; Susan B. Crockenberg and Barbara A. Soby, "Self-Esteem and Teenage Pregnancy"; and Harry H.L. Kitano, "Alcohol and Drug Use and Self-Esteem: A Sociocultural Perspective" – all in Mecca, Smelser, and Vasconcellos (eds.), *Social Importance of Self-Esteem*, 60, 135, 320.
30 https://www.theguardian.com/lifeandstyle/2017/jun/03/quasi-religious-great-self-esteem-con.
31 "Americans Seize the Chance to Feel Good about Doing Badly," *London Times* (April 7, 1990).
32 Nolan Jr., *The Therapeutic State*, 152–6.
33 "Schools' Responsibility for Low Self-Esteem" and "Letters to the Editor," *New York Times* (March 25, 1990).
34 "Down from the Self-Esteem High," *New York Times* (August 1, 1993).
35 "Self-Image is Suffering from Lack of Esteem," *New York Times* (May 5, 1998).
36 "Hey, I'm Terrific!," *Newsweek* (June 16, 1992).
37 "Teaching Self-Esteem Stirs Furor," *New York Times* (October 3, 1993).
38 "School Suspends Puppet Program," *New York Times* (February, 11 1992).
39 "Everett Schools Slay Self-Esteem Dragon," *Seattle Times* (September 18, 1992).
40 For an attempt to reconcile religious and secular perspectives, see Joanna McGrath and Alister McGrath, *The Cross and Christian Confidence* (Crossway Books, 1992).

41 "Teaching Self-Esteem Stirs Furor."
42 "Self-Esteem: Hoax or Reality?" *New York Times* (November 4, 1990).
43 Derek Colquhoun, "Researching with Young People on Health and Environment: The Politics of Self-Esteem and Stress," *Health Education Research*, 12 (1997), 449.
44 Mark R. Leary, "Making Sense of Self-Esteem," *Current Directions in Psychological Science*, 8 (1999), 333.
45 Kahne, "The Politics of Self-Esteem," 18.
46 See, among others, Alexander, *The New Jim Crow*; Anderson, *White Rage*, Ch. 4.
47 Roy F. Baumeister and Laura Smart, "Relation of Threatened Egotism to Violence and Aggression," *Psychological Review*, 103 (1996).
48 Brad J. Bushman and Roy F. Baumeister, "Threatened Egotism, Narcissism, Self-Esteem and Direct and Displaced Aggression: Does Self-Love or Self-Hate Lead to Violence?" *Journal of Personality and Social Psychology*, 75 (1998).
49 "You're OK, I'm Terrific. Self-Esteem Backfires," *Newsweek* (July 13, 1998).
50 Bernard Cornwell, "In America They Even Groom their Kids to Kill," *Daily Express* (April 22, 1999).
51 E.g. James P. McGee and Caren R. DeBernado, "The Classroom Avenger: A Behavioral Profile of School Based Shootings," *Forensic Examiner*, 8 (1999).
52 Brad J. Bushman, "Narcissism, Fame Seeking and Mass Shootings," *American Behavioral Scientist*, 62 (2017).

Chapter 9: The Twenty-first Century
1 Neil J. Mackinnon, *Self-Esteem and Beyond* (Palgrave, 2015), 6–11.
2 Franklin Holloway (ed.), *Self-Esteem: Perspectives Influences and Improvement Strategies* (Nova, 2016), vii.
3 David Miller and Teresa Moran, *Self-Esteem: A Guide for Teachers* (SAGE, 2012), 3–4; Mackinnon, *Self-Esteem and Beyond*, 2, 10, 15.
4 Polly Young-Eisendrath, *The Self-Esteem Trap: Raising Confident and Compassionate Kids in an Age of Self-Importance* (Little, Brown and Co., 2008), 4, 14, 21–2.
5 Vex King, *Good Vibes, Good Life: How Self-Love is the Key to Unlocking Your Greatness* (Hay House, 2018), xiv.
6 Mary H. Guindon (ed.), *Self-Esteem Across the Lifespan* (Routledge, 2010), xii.
7 Eva Illouz, *Why Love Hurts: A Sociological Explanation* (Polity, 2012), 13.

8 Richard L. Allen, *The Concept of Self: A Study of Black Identity and Self-Esteem* (Wayne Street University Press, 2001).
9 Tarana Burke and Brené Brown, *You Are the Best Thing: Vulnerability Shame Resilience and the Black Experience* (Vermilion, 2021).
10 Dyson, *Pride*, xiv–xv, 25.
11 hooks, *Rock My Soul*, 4, 44.
12 Glaude Jr., *Democracy in Black*, 23–4.
13 Taylor, *From #Blacklivesmatter to Black Liberation*, 24–5.
14 Shelby Steele, *The Content of our Character: A New Vision of Race in America* (St. Martin's Press, 1990); James L. Robinson, *Racism or Attitude? The Ongoing Struggle for Black Liberation and Self-Esteem* (Insight Books, 1995); John H. McWhorter, *Losing the Race: Self-Sabotage in Black America* (Free Press, 2000).
15 Barbara Herlihy and Zarus E.P. Watson, "Self-Esteem of African American Adolescent Girls," in Guindon (ed.), *Self-Esteem Across the Lifespan*, 174.
16 E.g. Kelly Huegel Madrone, *LGBTQ: The Survival Guide for Lesbian, Gay, Bisexual, Transgender and Questioning Teens* (Free Spirit, 2018); Fox Fisher and Owl Fisher, *Trans Teen Survival Guide* (Jessica Kingsley, 2018); Owl Fisher and Fox Fisher, *Trans Survival Workbook* (Jessica Kingsley, 2021).
17 Richard Ruth and Erik Santacruz (eds.), *LGBT Psychology and Mental Health: Emerging Research and Advances* (Praeger, 2017).
18 Odets, *Out of the Shadows*, 73.
19 Downs, *The Velvet Rage*, x.
20 https://www.psychologytoday.com/gb/blog/how-raise-happy-cooperative-child/201205/simple-ways-build-your-babys-self-esteem.
21 https://www.askdrsears.com/topics/parenting/child-rearing-and-development/12-ways-help-your-child-build-self-confidence/.
22 Tara Losquadro Liddle, *Why Motor Skills Matter: Improve Your Child's Physical Development to Enhance Learning and Self-Esteem* (Contemporary Books, 2004), xx.
23 Shani Orgad and Rosalind Gill, *Confidence Culture* (Duke University Press, 2022), 100–23.
24 Moira McCarthy, *The Everything Guide to Raising Adolescent Girls* (Adams Media, 2008), 77–90.
25 Erika V. Shearin Karres, *The Everything Parent's Guide to Raising Girls*, 2nd ed. (Adams Media, 2011 [2007]), 188–201.
26 Beverly Engel, *Healing Your Emotional Self* (Wiley, 2006), 1.
27 E.g. Carolyn M. Lawrence, *The Negative Impacts of Politics on Literacy: The Importance of Self-Esteem for Reading Achievement*, 2nd ed. (Rowman and Littlefield, 2014).

28 Miller and Moran, *Self-Esteem*, 3.
29 C. Bobbi Hansen, *The Heart and Science of Teaching* (Teacher's College Press, 2019), 12–13.
30 National Commission on Social, Emotional and Academic Development, *From a Nation at Risk to a Nation at Hope* (Aspen Institute, 2019).
31 E.g. Naidi Okeke-Adeyanju, Lorraine C. Taylor, Ashley B. Craig, et al., "Celebrating the Strengths of Black Youth: Increasing Self-Esteem and Implications for Prevention," *Journal of Primary Prevention*, 35 (2015).
32 Nancy Evans, *Cultivating Strong Girls: Library Programming that Builds Self-Esteem and Challenges Inequality* (Libraries Unlimited, 2018).
33 Katty Kay and Claire Shipman, *The Confidence Code for Girls: Taking Risks, Messing Up and Becoming your Amazingly Imperfect, Totally Powerful Self* (HarperCollins, 2018); Lisa Hinkelmann, *Girls without Limits: Helping Girls Succeed in Relationships, Academics, Careers, and Life* (Corwin Press, 2021).
34 E.g. Pat Palmer and Melissa Alberti Froehner, *Teen Esteem: A Self-Direction Manual for Young Adults*, 2nd ed. (Impact, 2000 [1989]).
35 Orgad and Gill, *Confidence Culture*, 2.
36 Mark J. Warner, *The Complete Idiot's Guide to Enhancing Self-Esteem* (Alpha Books, 1999); S. Renee Smith and Vivian Harte, *Self-Esteem for Dummies* (Wiley, 2015).
37 Gael Lindenfield, *Self-Esteem: Simple Steps to Build your Confidence* (HarperCollins, 2014 [1995]), x.
38 E.g. Kate Collins-Donnelly, *Banish your Self-Esteem Thief: A Cognitive Behavioral Therapy Workbook on Building Positive Self-Esteem for Young People* (Jessica Kingsley, 2014).
39 Judith Belmont, *Embrace your Greatness* (New Harbinger, 2019), 1–2.
40 E.g. Glenn R. Schiraldi, *Ten Simple Solutions for Building Self-Esteem* (New Harbinger, 2006).
41 Belmont, *Embrace your Greatness*, 11–13.
42 Duckworth, *Grit*.
43 Eileen Kennedy-Moore, *Kid Confidence: Help Your Child Make Friends, Build Resilience and Develop Real Self-Esteem* (New Harbinger, 2019), 96.
44 Heidi A. Wayment and Jack J. Bauer (eds.), *Transcending Self-Interest: Psychological Explorations of the Quiet Ego* (American Psychological Association, 2008).
45 See, among others, Martin E.P. Seligman, *Learned Optimism: How to Change your Mind and Your Life*, 2nd ed. (Simon and Schuster, 2005 [1991]); *Authentic Happiness: Using the New Positive Psychology to Realize Your Potential for Lasting Fulfilment* (Free Press, 2004); *Flourish: A Visionary New Understanding*

of Happiness and Wellbeing (Free Press, 2011); Christopher Peterson and Martin Seligman, *Character Strengths and Virtues* (Oxford University Press, 2004).
46 Horowitz, *Happier?*, 4, 66.
47 Miller and Moran, *Self-Esteem*, 5.

Index

Abernethy, Virginia 164
Ablon, Steven, and Mack, John 113, 116, 154
Adams, Gerald R., and Crossman, Sharyn M. 178
Adler, Alfred 29–30, 31, 38–9, 43–4, 50, 58, 213
　see also inferiority, and inferiority complex; psychoanalysis
adolescence 9, 116, 127, 150, 151–66, 167, 220, 221
　and girls 167, 185–96, 220, 221
　and pregnancy 115, 139, 153, 158, 164–5, 201, 202, 204, 206
　and schools 109, 185–96
　see also schools
African Americans 1, 6, 16, 17, 18, 19, 46, 47–75, 76, 78, 100, 102, 106, 173, 197, 214, 216, 218, 224
　and children 54, 71, 105, 136
　and crime 56, 59, 73–4, 210–12
　and families 53, 60–1, 103, 117–27, 145, 148–9
　and film 68
　and hair 56, 64, 65–7
　and LGBTQ+ 18–19, 89–91, 93, 94
　and literature 67–8
　and music 67
　and passing 53–4
　and schools 17, 55–6, 128–9, 137, 141–50
　and sport 68
　and teenagers 153, 166
　and women 18–19, 57, 59, 68–9, 117–27, 167, 194–6
　and work 56
　see also Black Panther Party; Black Power; Black pride; civil rights movement; inferiority and African Americans
AIDS 17, 76, 77, 91, 97–102, 204, 206
alcohol 14, 36, 38, 95, 105, 113, 115, 136, 154, 158, 160, 162, 165, 170, 189, 200, 203, 205–6
Ali, Muhammed 68
alienation 3, 29, 97, 147, 157, 160
ambition 21
American Association of University Women 185–90
American Psychiatric Association 8, 92
American Psychological Association 32
anomie 29
anthropology 22
anxiety 1, 3, 4–5, 11, 12, 13, 31, 40, 139, 152, 157, 161, 215, 222
Asian Americans 19
Association for the Advancement of Behavior Therapy 93

Badinter, Elizabeth 107
Baldwin, James 51, 67–8, 69
Banks, James A. 142–3
Baughman, Jr., E. Earl, 47, 62, 72
Baumeister, Roy F. 165, 210–11
Beane, James, and Lipka, Richard 138
Beatles 41, 42, 45–6
Becker, Ernest 44–5
Bell, Derrick 144–5
Bergler, Edmund 77–8, 87
Berzon, Betty, and Leighton, Robert 93

biculturality 121
Bieber, Irving 87
Billingsley, Andrew 119
bipolar disorder 3
Black Panther Party 18, 47, 61–5, 68
  see also Black Power
Black Power 17, 18, 47–8, 59–65, 69, 74, 142, 199
  see also Black Panther Party
Black pride 18, 19, 58–60, 62, 71, 142
Blake, Roger 78
Blaker, Karen 182
Borba, Michele and Craig 128–9, 130, 137
Bowlby, John 110
Boyd, Julia 4, 57, 194
Boyd-Franklin, Nancy, and Franklin, Anderson J. 103, 105, 126
Branden, Nathaniel 44–5, 136, 169
Brando, Marlon 36–7, 39
Briggs, Dorothy Corkville 112, 114–16, 136
Brown, James 67
Burns, Robert R. 134

Calvinism 40
Canfield, Jack 14–15, 136, 202
capitalism and neoliberalism 2, 3, 11, 12–13, 19, 41, 46, 47, 75, 77, 103, 106, 114, 121, 129, 131, 132, 153, 167, 171, 176, 185, 193, 198, 201, 213–14, 216, 219
Carmichael, Stokely 61–3
Carnegie, Dale 34
Cass, Vivienne 97
Chapkis, Wendy 183
character 21, 24, 27, 32, 37
Chavez, Brian 86
child abuse 14, 105
children (general) 38, 40, 44–5, 105, 114, 128–50, 152
  and infants 105–27
China 3

civil rights movement 47, 54, 55, 56, 59, 61, 62, 63–4, 66, 67, 69–70, 71, 74, 91, 103, 104, 107, 153
Clark, Don 94–5
Clark, Kenneth B. 55–6, 57, 69–70, 117
Clarke, Jean Illsley 106–7, 113–14, 116–17
class 19, 49, 54–6, 62, 105, 106, 111, 117–18, 120, 122, 124, 131, 153, 155, 159, 160, 165–6, 171, 194, 219, 224
Cleaver, Eldridge 68–9
Coles, Robert 54, 144
Colman, Carol, and Eagle, Carol 170
Comer, James, and Pouissant, Alvin 123, 147–8
competitiveness 3, 21
conservativism 12, 13, 19, 77, 103–4, 129–30, 151, 157, 208–9, 214
  see also Schlafly, Phyllis
Cooley, Charles H. 29
Coopersmith, Stanley C. 8, 110
counterculture 41–6, 48, 70, 103, 104, 126
Cory, Donald Webster 82–3, 86
COVID 5, 17, 47, 75
crime 26, 38, 105, 109, 139
Crossman, Sharyn M., and Gerald R. Adams, 178
cults 158, 159

Darwin, Charles 26
Daughters of Bilitis 81, 89–90
Dean, James 36, 39
degeneration 26
depression 3, 5, 8, 11, 14, 93, 113, 152, 157, 171, 186, 220, 222
despair 12, 13, 77, 98, 138
Dickens, Charles 21
dieting 179–80, 184
Donovan, Mary Ellen, and Sanford, Linda Tschirhart 172
Douglas, Susan J. 177
Dr. Dre 210

INDEX

drugs (illegal) 14, 15, 95, 113, 115, 136, 139, 152, 153, 154, 157, 158, 161–3, 165, 170, 189, 193, 202, 203, 205–6, 210
Du Bois, W.E.B. 50, 58
Dungeons and Dragons 158
Duso the Dolphin 207
Dyson, Michael Eric 128, 129, 131, 217

Eagle, Carol, and Colman, Carol 170
eating disorders 3, 14, 170, 179–80
educational psychology 130–1, 131–41
Eichberg, Rob 93
Ellis, Albert 87
Ellison, Ralph 52
emotions theories 25–6
English, Deirdre 175–6
envy 21, 24
epidemiology (of low self-esteem) 1, 2–3, 5, 33, 70–2, 150–1
Erikson, Erik 109, 154
Esalen Institute 45, 198
eugenics 26, 27, 50, 51, 110
extroversion 4, 9

Faludi, Susan 176, 183
Family Research Council 202
Fanon, Frantz 58, 69
Fat Liberation Movement 182
Faust, Verne 139
feminism 1, 16, 17, 18, 19, 103, 107, 110, 166, 167–96, 197, 199, 214, 218
  and African Americans 18, 168
Floyd, George 75
Franklin, Anderson J., and Boyd-Franklin, Nancy 103, 105, 126
Frazier, E. Franklin 62
Freedman, Rita 179, 180
Freud, Anna 36, 111, 154
Freud, Sigmund 6, 26–7, 30, 42, 58, 88–9, 96, 109, 130, 213
  see also psychoanalysis

Friedan, Betty 37, 169
Fromm, Erich 37
Frye, Marquette 118–19
Fuller, Marielle, and Powell, Gloria J. 121–2

Gall, Franz Josef 25
Garland, Judy 36
Garvey, Marcus 60
gay pride 17, 18, 76, 85, 86, 88, 91, 97, 98, 101–2, 103, 218
Gay Sunshine 86
gender (general) 17, 19, 34, 105, 130
Geraty, Ronald 129
Gestalt Therapy, 45
Gilligan, Carol 186–7
girls 17, 57, 105, 128, 143–4, 164–5, 167, 185–96, 221
  see also adolescence; feminism; women
Goode, Starr 168–9
grief 17, 98, 222
grit 13, 223
guilt 14, 74, 83

Hacking, Ian 9, 35
Hale, Janice E. 149
Hall, Granville Stanley 153–4
Hamilton, Charles 62
Hancock, Emily 187
happiness 3, 5, 9, 19, 21–3, 27, 33, 40, 103, 105, 160, 165, 181, 192, 215
Harper, Frederick D. 147
Harper, Robert Allen 79–81
Harris, Patricia 118
Harrison-Ross, Phyllis, and Wyden, Barbara 124
Hatterer, Lawrence J. 87
Hay, Harry 82
  see also homophile movement; Mattachine Society
Hayworth, Rita 36, 39

271

INDEX

heavy metal 158, 160
Hill, Napoleon 34
Hillman, Carolynn 174–5
Hinduism 41, 209
hip-hop 210–11
Hispanics 19
Holloway, Franklin 215
homophile movement 76, 80–7
Hooker, Evelyn 94
hooks, bell 17, 69, 73, 74, 195–6, 217
Horney, Karen 30–3, 39, 40, 45, 116
Houston, Whitney 223
Human Potential Movement 42–3, 45–6, 65, 75, 93
Huxley, Aldous 42

imposter syndrome 170
incest 14
individualism 9, 27, 104, 121, 176, 198, 216, 219
inferiority 3, 5, 6–7, 165
    and African Americans 48, 49–52, 56, 57, 58, 61, 64, 72, 73, 74
    and inferiority complex 29–30, 32, 33, 36–7, 38–9, 56, 57, 58, 61
    see also Adler, Alfred
Isay, Richard A. 96

Jackson, Jesse 72
James, William 6, 28, 29
Janov, Arthur 46
    see also primal scream
Johnson, Lyndon B. 136
Johnson, Robert C. 125–6
Jones, Silas 24

Kaplan, Howard B. 156, 158
Kardiner, Abram, and Ovesey, Lionel 52, 53–4, 57, 60, 70, 72
Keil, Charles 67
Kennedy, John F. 204
King, Martin Luther 61, 62, 63, 119
Kinsey, Alfred 83

Krate, Ronald, and Silverstein, Barry 120–1
Kunjufu, Jawanza 148–9

Ladner, Joyce A. 120
Lasky, Ella 178
Latinos 19
Leary, Timothy 41–2
Leighton, Robert, and Betty Berzon 93
Leonetti, Robert 134–5
lesbians 81, 83, 88–90, 91, 93, 94, 97, 98, 102, 167, 214
    see also Daughters of Bilitis
LGBTQ+ 1, 6, 8, 16, 17, 18, 19, 46, 75, 76–102, 126, 153, 166, 167, 173, 197, 199, 212, 214, 216, 224
    and African Americans 18–19, 89–91, 93, 94, 100, 102, 218
    and coming out 96–7
    and families 78
    and Gay Liberation Front 87
    and gay press 84–5, 85–7, 93
    and homophile movement 75, 81–9
    and pathologization of 8, 9, 77–81, 85–6, 92, 96
    and positive therapies 92–7
    and Stonewall 77
    see also AIDS; Daughters of Bilitis; gay pride; lesbians; Mattachine Society; Society for Individual Rights (SIR)
Lipka, Richard, and James Beane 138
Lipton, Lawrence 42
Logan, Joshua 36
loneliness 3, 4, 5, 14, 36, 40, 42, 77, 138, 152
    see also social isolation
LSD 41, 42, 45
Lyon, Phyllis 89–90

Mack, John, and Ablon, Steven 113, 116, 154

Maharishi 42
Malcolm X 59–61, 64, 66, 69, 147
Martin, Del 89–90
Maslow, Abraham 4, 43–4, 45, 169
    see also needs (hierarchy of)
masturbation 88
Mattachine Society 81–2, 84, 85, 91
Mayo, Mary Ann 180–1
Mead, George Herbert 28–9, 30
Mecca, Andrew 200, 203, 205
meditation 41, 42, 45, 207
meritocracy 11
Methodism 22
migration 3
Miller, Merle Dale 77, 86
Millett, Kate 169
mindfulness 12–13, 214
Miss America 177
modernization 2, 3, 5
modesty 25
Monroe, Marilyn 36, 39
Morrison, Toni 143–4
Moynihan, Daniel Patrick 117–18, 119

narcissism 10, 25–6, 151–2, 177, 202, 213, 222
National Association of Secondary School Principals 158
National Institute of Mental Health 32, 155
National Urban League 119
Native Americans 19
needs (hierarchy of) 4, 43–4
    see also Maslow, Abraham
Nelson, Gene 36
neurosis 31, 79, 80, 114
New Age 93, 209
Nixon, Richard 104
normality 19, 32, 33, 37–8, 103, 105, 116, 127, 150, 152, 156, 166, 219
Nozick, Robert 199

Odets, Walt 17, 99–100, 101–2

O'Killens, John 64–5
Orbach, Susie 179–80
Orenstein, Peggy 187–8, 189
Ovesey, Lionel 87
    and Kardiner, Abram 52, 53–4, 57, 60, 70, 72
Owen, Robert 24–5

parenting 4, 15, 103–27, 151, 152, 153, 166, 170, 200, 214, 219
Parks, Rosa 59, 64, 68
    see also civil rights movement
Patterson, Cecil H. 132
Peale, Norman Vincent 34–5
perfection 11
Pettigrew, Thomas 54
phrenology 24–5
physical appearance 33, 34, 125, 151, 155, 220
    and beauty 167, 177–85
    and men 33
    and women 9, 33
    see also plastic surgery
physical health 1, 3, 14, 47, 57, 91, 95, 106, 109, 113, 120, 139, 154, 198
    see also AIDS
Piaget, Jean 128
Pinker, Steven 191
Pinney, Morgan 86
Pipher, Mary 170–1, 173, 180–1, 187, 188, 193–4
plastic surgery 167, 181–2, 183–4
Poitier, Sidney 68
Popper, Karl 199
positive psychology 13, 26, 223–4
positive thinking 21, 34, 35–7, 40
Pouissant, Alvin 62, 63–4, 122–3
    and Comer, James 123, 147–8
Powell, Gloria J., and Fuller, Marielle 121–2
Powers, Melvin 34
Presley, Elvis 36, 39, 182
Presley, Priscilla 182

pride 6, 7, 23–5
  see also Black pride; gay pride
primal scream 45
  see also Janov, Arthur
Proud Boys 19
psychoanalysis (general) 25–6, 26–7, 36, 40, 42–3, 92–3, 207
Pumsy the Dragon 207–8
Purkey, William Watson 134

race (general) 16, 19, 47–75, 105, 106, 131, 134, 138–9, 141–50, 204, 210–12, 216–18, 220, 224
Racial Identity Attitude Scale (RIAS) 94
Rainwater, Lee 120
Rands, William Brighty 24–5
rape 14, 15, 59, 69, 158, 187
Reagan, Ronald 12, 145, 162
rejection 14
religion 4, 5, 7, 11, 14, 23, 34–5, 84, 91, 100, 158, 165, 198, 207–9
resilience 13, 190
Rogers, Carl 42–3, 43–4, 45, 169
Rome, Esther, and Wegscheider, Jane 182
Rose, Arnold M. 58
Rose, Nikolas 6
Rosenberg, Morris 7, 38, 71, 111, 119–20, 154–5
Ross, Diana 67
Ross Laboratories 109

Sadker, Myra and David 186, 192
Samuels, Shirley C. 111, 112–13
Sanford, Linda Tschirhart and Donovan, Mary Ellen 172
Satanism 153, 158–60, 164, 165
Schlafly, Phyllis 18–19, 103–4, 107, 208–9, 223
schizophrenia 107
schools 128–50, 156–7, 220–1
  and academic performance 14, 55–6, 105, 131, 133, 157, 185–96, 206
  and educational strategies 14, 15, 19, 22, 131–41, 205
  and kindergarten 128, 166
  and shootings 210–11
  and teachers 109, 114, 128, 132–3, 133–41, 142–3, 144–5, 151, 152, 185–90, 214, 219
  see also adolescence, and schools; African Americans, and schools; educational psychology
self-actualization 22, 43–5, 156
self-care 17, 18
self-confidence 4, 6, 9, 19, 21, 22, 34, 39, 49, 58, 59, 75, 82, 109, 115, 137, 138, 139, 151, 153, 162, 165, 169, 172–3, 176, 178, 192, 193, 198
self-esteem, definitions 7–10
self-harm 154, 170
self-help (general) 19, 21–2, 32–3, 37, 64, 75, 82, 103, 104–5, 131, 157, 158, 167, 168, 171, 175, 187, 193, 212, 213–14, 215, 218
  and literature 1, 4, 7, 10, 13, 14–15, 19, 20–21, 34–5, 76, 110–17, 139, 187, 215, 222–3, 224
  and classes, workshops, and courses 10, 14, 41, 64
self-importance 25
selfishness 21, 25
Sellers, Sean 159
sexuality (general) 16, 19, 76–102, 105
Shaft, John 68
shame 17, 74, 76, 77, 82–3, 84, 98, 101, 102, 215, 216
Shannahoff-Khalsa, David 203, 206
Sherman, Vincent 36
Silverstein, Barry, and Krate, Ronald 120–1
Silverstein, Charles 92–3
Simpson, George Eaton, and Yinger, J. Milton 58
Sizer, Theodore and Nancy 133–4
Smesler, Neil J. 205–6

Smith, Donald H. 141–2
smoking 14, 154, 162, 163–4
Snoop Doggy Dogg 210
Socarides, Charles W. 78, 87
Social Darwinism 26
social isolation 2, 3, 4, 12, 33, 77, 97,134
  see also loneliness
social (in)justice 12, 16, 19, 75, 131, 160, 214
social media 11
Society for Individual Rights (SIR) 84, 85
Sommers, Christina Hoff 189–90
Spock, Benjamin 111–12, 128
Steinem, Gloria 17, 18–19, 167, 173–4, 175–6, 187, 188, 196
Stetler, Stacey 179
stigma 17
Stonequist, Everett V. 52–3, 62
stress 11, 222
success (economic) 21, 34, 35, 57
suicide 36, 89, 139, 152, 154, 157, 165, 171, 189, 220
Sullivan, Harry Stack 30–1, 109
  see also psychoanalysis
Szasz, Thomas 8, 12

tai-chi 45
Task Force to Promote Self-Esteem and Personal and Social Responsibility 197–206, 222
  see also Vasconcellos, John
Taylor, Elizabeth 184
Temple, Shirley 144
Terrell, Mary Church 50–1
therapeutic culture 11–12, 15–20, 47, 51, 130, 134–5, 157, 167
Toddler Infant Experiences Survey 125
trauma 8, 13, 17, 118, 186
Tripp, Clarence A. 92
Trudeau, Gary 201, 202
Turner, Tina 182

unconscious data bias 50
urbanization 2, 3

Vasconcellos, John 197–99, 201–6, 212, 222
"vice patrol" approach 79
violent crime 14, 15, 41, 59, 69–70, 118, 142–3, 158, 159, 171, 182, 200–1, 203, 210–11
Vitz, Paul C. 7

Washington, Booker T. 58
Wegscheider, Jane, and Rome, Esther 182
Weinberg, George 96
wellbeing and wellness 1, 17, 68, 214
West, Donald J. 83
White, Peggy 168–9
white supremacy 17, 19, 54–5, 59–60, 73, 104, 129, 141–2
Williams, A. Cecil 84, 91
Wilson, Amos N. 124–5
Winters, Mary-Frances 74, 128, 129, 131, 144–5
witchcraft 158
Wolf, Naomi 179
women 17, 18, 19, 33, 35–6, 41, 78, 79, 88–90, 105, 167–96, 217, 221–2
  see also African Americans, and women; feminism; gender; girls; lesbians
work 17, 21, 39, 56, 171–2, 186, 222
Wright, Richard 51, 59
Wyden, Barbara, and Harrison-Ross, Phyllis 124
Wylie, Ruth 169, 188

Yinger, J. Milton, and Simpson, George Eaton 58
yoga, 45

Zigler, Edward 136